MATHS IN ACTION

ADVANCED
HIGHER

Mathematics 1

Edward Mullan
William Richardson
Clive Chambers

This edition first published in 2000 by:
Thomas Nelson and Sons Ltd

Reprinted in 2001 by:
Nelson Thornes Ltd
Delta Place
27 Bath Road
CHELTENHAM
GL53 7TH
United Kingdom

09 10 11 12 / 13 12 11 10

A catalogue record for this book is available from the British Library

ISBN 978 0 17 431541 4

Picture Research by Zooid Pictures Limited
Typeset by Upstream, London

Acknowledgements
The authors abd publishers are grateful for permission to include the following copyright material:
Corbis UK Ltd; p. 1 (Archivo Iconographico SA), p. 25 (top left), p. 68 (Leonard de Selva): Science Photo Library; p. 25 (bottom left), p. 25 (right – Dr Jeremy Burgess), p. 50 (George Bernard), p. 96 (National Library of Medicine), p. 127 (Karl Friedrich Gauss).

Printed by Multivista Global Ltd

$Contents$

Preface

This book is part of the *Maths in Action* series, and has been written to address the needs of students following Advanced Higher Mathematics, unit 1, Mathematics 1. It is assumed that students have already done Higher Mathematics, and each chapter contains some reminders of necessary knowledge.

The order of the chapters and the topics within chapters mirror that in the detailed contents list of the national course specifications. However, certain items appear out of order within the text because their use is required before they are formally introduced. For example,

- $|x|$ is required for integrating x^{-1} (Chapter 3) before it gets a formal mention in Chapter 4 on Properties of Functions;
- concavity is required if the second derivative is to be used to test the nature of stationary points, before its formal exposition in Chapter 4;
- the main handling of matrices will be in *Mathematics 3* but they are informally introduced in this book so that they can be used for Gaussian elimination.

Anomalies of this nature will always occur in a course when it is unitised.

Notation and conventions follow those of the book *Higher Mathematics, Maths in Action*. The style also reflects earlier books in the *Maths in Action* series so that students can start this course with as seamless a join as possible. Some liberties have been taken in mixing language and notation to ease the student into more rigorous thought as gently as possible. Explanations have been kept short and to the point to ensure that the topics remain accessible.

Minor historical notes have been added at places throughout the text in the hope that students are made curious enough to pursue the topics on their own. The historical background to a piece of maths (dates, personalities, inspiration and development) often helps to put it in a context.

Each chapter ends with two features:

- a Review exercise which has the purpose of ensuring the student that he or she has indeed picked up all the learning outcomes associated with that particular chapter;
- a Summary which provides the basis of revision notes.

1.1 Factorials and the Binomial Theorem

Pascal's triangle

Historical note

Blaise Pascal

Blaise Pascal was a French mathematician born in 1623. He invented one of the earliest calculating machines and developed much new material.

In 1654, an expert gambler, the Chevalier de Méré, consulted Pascal about certain problems to do with games of chance. Pascal wrote to his friend, Pierre De Fermat, about the problems and between them they produced what was to become the theory of probability. It was while working on this that he produced a treatise called *Traité du triangle arithmétique* in which he discussed a number pattern which now bears his name.

EXERCISE 1

1 When two coins are tossed, we can use a tree diagram to help us enumerate the possible outcomes.

A summary of the four outcomes

Number of Heads	0	1	2
Frequency	**1**	**2**	**1**

a Construct a tree diagram to help you complete the table which summarises the eight possible outcomes when three coins are tossed.

Number of Heads	0	1	2	3
Frequency	**1**			

b Complete the table which summarises the 16 possible outcomes when four coins are tossed.

Number of Heads	0	1	2	3	4
Frequency	**1**				

2 $(x + y)^2 = x^2 + 2xy + y^2$

This multiplication can be represented as in Table 1 or more compactly, using coefficients only, as in Table 2.

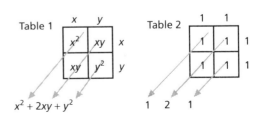

1

For $(x + y)^3 = (x + y)(x + y)^2 = (x + y)(x^2 + 2xy + y^2)$

A similar compact table produces

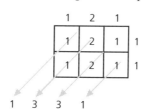

> **Note**
> • the descending powers of x
> • the ascending powers of y

$$(x + y)^3 = x^3 + 3x^2y + 3xy^2 + y^3$$

a Use this technique to expand **(i)** $(x + y)^4$ **(ii)** $(x + y)^5$ **(iii)** $(x + y)^6$.

b Compare these results with the tables of question 1.

3 The coefficients of these expansions, arranged in order, form the number pattern known as Pascal's triangle (see below).

Each entry in the triangle is the sum of two entries above it: the one immediately above it, and the one above it and to the left.

This spreadsheet has been created to build Pascal's triangle:

1 • Enter 0 in B1.
  • Enter the formula '= B1 +1' in C1.
  • Fill right.
2 • Enter 0 in A2.
  • Enter the formula '= A2 +1' in A3.
  • Fill down.

This row and column will be used as references and don't form part of the triangle. They will be useful when the pattern is not in the context of a spreadsheet.

3 • Enter 1 in B2.
  • Fill down.
4 • Enter the formula '= B2 + C2' in C3.
  • Fill down.
  • Highlight all of C3 to I14 and fill right.

5 Delete zero entries to highlight the triangular pattern.

	A	B	C	D	E	F	G	H	I
1		0	1	2	3	4	5	6	7
2	0	1							
3	1								

		0	1	2	3	4	5	6	7
4	0	1							
5	1	1	1						
6	2	1	2	1					
7	3	1	3	3	1				
4		1	4	6	4	1			
5		1	5	10	10	5	1		
6		1	6	15	20	15	6	1	
7		1	7	21	35	35	21	7	1
8		1	8	28	56	70	56	28	8
9		1	9	36	84	126	126	84	36
10		1	10	45	120	210	252	210	120
11		1	11	55	165	330	462	462	330
12		1	12	66	220	495	792	924	792

a For the pattern in column 0, $u_n = 1$ where n is the row number. For the pattern in column 1, $u_n = n$. Find a formula for **(i)** column 2 **(ii)** column 3

b The coefficients of the expansion $(x + y)^n$ are to be found in row n. By examining the symmetry in the rows, write down the expansion of
  (i) $(x + y)^8$ **(ii)** $(x + y)^9$ **(iii)** $(x + y)^{10}$

c Each entry can be referenced using a number pair given in the format (row, column), for example (5, 3) is the entry 10 in the 5th row, 3rd column
  (4, 2) is the entry 6 in the 4th row, 2nd column

Traditionally we refer to row n and column r: (n, r).
  (i) Identify the entry in $(5, 0)$, $(5, 1)$, $(5, 2)$, $(5, 3)$, $(5, 4)$, $(5, 5)$.
    Comment on the symmetry by completing the equation $(n, r) = \ldots$
  (ii) In general, is entry $(a, b) = (b, a)$?
    Note that the property used to *build* the triangle is that
    $(n, r - 1) + (n, r) = (n + 1, r)$

d There are many patterns to be found in the triangle.
  (i) Why should the sum of any two adjacent numbers in column 2 always be a perfect square?
  (ii) Diagonals which slope up to the right can be identified by giving the row reference in which they start. Explore the *sum* of their entries. For example:

    diagonal 1: 1　　　　 = 1
    diagonal 2: 1　　　　 = 1
    diagonal 3: 1 + 1　　 = 2
    diagonal 4: 1 + 2　　 = 3
    diagonal 5: 1 + 3 + 1 = 5
    diagonal 6: 1 + 4 + 3 = 8

> These are known as the Fibonacci numbers.
> If F_n is the nth Fibonacci number, investigate the quotient $F_{n+1} \div F_n$.

  Look for more patterns!

e Isaac Newton explored the development of the triangle *upwards* to study the expansion of expressions of the form $(x + y)^{-n}$. Investigate such an upward development.

The factorial function

Example 1　Six cyclists enter a race. All six finish at different times. There is a 1st and a 2nd prize.
a How many different ways can the cyclists finish the race?
b How many different ways can the prizes be awarded?

a There are 6 different possible winners.
  For each possibility there are 5 possible *seconds*: $6 \times 5 = 30$ ways of getting a 1st and a 2nd.
  For each of these, there are 4 possible *thirds*: $6 \times 5 \times 4 = 120$ ways of getting a 1st, 2nd and 3rd.
  For each of these, there are 3 possible *fourths*: $6 \times 5 \times 4 \times 3 = 360$.
  For each of these, there are 2 possible *fifths*: $6 \times 5 \times 4 \times 3 \times 2 = 720$.
  For each of these, there is only 1 possible *last*: $6 \times 5 \times 4 \times 3 \times 2 \times 1 = 720$.
  Therefore there are 720 different ways the cyclists can finish the race.
b Looking at the working above we see that there are $6 \times 5 = 30$ ways of getting a 1st and a 2nd prize.

Calculations similar to $6 \times 5 \times 4 \times 3 \times 2 \times 1 = 720$ appear often enough for there to be a special function created, to represent them. $n!$, read as n *factorial*, is defined as $n \times (n - 1) \times (n - 2) \times \ldots \times 3 \times 2 \times 1$. For example, $6! = 6 \times 5 \times 4 \times 3 \times 2 \times 1 = 720$

Most calculators carry this function. $\boxed{n!}$

Part **b** of Problem 1 can be answered using factorials when you note that

$$6 \times 5 = \frac{6 \times 5 \times 4 \times 3 \times 2 \times 1}{4 \times 3 \times 2 \times 1} = \frac{6!}{4!}$$

This calculation is a special function ^nP_r and is also supported by most calculators: look for $\boxed{^nP_r}$. It calculates the number of ways of arranging r objects when they are first to be selected from a pool of n objects.

Example **1b** required the number of ways of arranging 2 cyclists when they are first to be selected from a pool of 6 cyclists.

$$^6P_2 = 6! \div (6 - 2)! = 30$$

$$^nP_r = \frac{n!}{(n-r)!}$$

P refers to the word *Permutation* which means *arrangement*.

Example 2 From a palette of 7 colours you can pick 4 to blend. How many ways can this be done?

The order in which the colours are blended does not matter: *red, blue, yellow, green* is the same choice as *red, blue, green, yellow*.

So, in this example, we are selecting 4 colours from a pool of 7 but the arrangement (order) of the 4 does not matter.

There are $4 \times 3 \times 2 \times 1 = 4! = 24$ ways of arranging 4 colours, so $^7P_4$, which includes all of these ways, will be 24 times too big.

We get our desired answer by performing the calculation $^7P_4 \div 4! = 35$

This also is a very common type of problem and this calculation has a special function ^nC_r, supported by most calculators:

look for $\boxed{^nC_r}$.

$$^nC_r = \frac{n!}{r!(n-r)!}$$

C refers to the word *Combination* which is *selection without arrangement*.

Example 3 *Hobson's Choice*. Hobson hired out horses. You paid your money and got the horse of your choice – as long as you chose the horse he offered you. How many ways can you pick one horse when there is only one horse from which to pick?

$$^1C_1 = \frac{1!}{1!(1-1)!} = \frac{1!}{1! \times 0!}$$

We know the answer is that there is only one way, so $\dfrac{1!}{1! \times 0!} = 1$

For this to make sense we must give a value of 1 to $0!$.

As a definition, we have: $0! = 1$.

Note
- the domain of the factorial function is W, the set of whole numbers.
- $n!$, ^nP_r and ^nC_r on a graphics calculator are generally found in the MATHS menu.

EXERCISE 2A

1 a Use your calculator to obtain values for **(i)** 4! **(ii)** 6! **(iii)** 0!
   b What happens when you attempt to get **(i)** −4! **(ii)** (−4)! **(iii)** 4.2!

2 a The combination on the lock on my case uses the four digits 1, 2, 3 and 4. I've forgotton the order in which they appear. How many different ways can they be arranged?
   b In a game of *Scrabble* a player has 7 different letters. He rearranges them, looking for words. How many different arrangements can he make of the 7 letters?
   c A pack of cards has 52 different cards. Calculate the number of ways these can be arranged, giving your answer correct to three significant figures.

3 a Evaluate **(i)** $\dfrac{7!}{3!}$ **(ii)** $\dfrac{10!}{6!}$ **(iii)** $\dfrac{12!}{11!}$
   b Can you account for the fact that the values of these divisions are all exact?
   c Use the ^nP_r facility on your calculator to evaluate the same expressions.

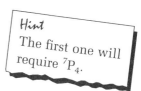

Hint
The first one will require $^7P_4$.

4 a From a class of 23 students, 3 have to be selected to be class representative, secretary and treasurer of the newly formed student committee. In how many different ways can this be done?
   b At the bank, a customer invents her own personal identification number (PIN) by choosing four different digits. Given that there are 10 different digits, how many arrangements of 4 different digits can be made? An arrangement may start with zero.
   c A driver has 5 tyres on his car: the 4 on the road and 1 in the boot. He rotates them regularly so that they wear evenly.

   (i) How many different arrangements of 4 tyres on the road can he make, assuming position matters?
   (ii) How many different arrangements of one tyre in the boot can he make?
   (iii) Comment on your answers.

5 a Evaluate **(i)** $\dfrac{7!}{4!3!}$ **(ii)** $\dfrac{10!}{6!4!}$ **(iii)** $\dfrac{12!}{11!1!}$ **(iv)** $\dfrac{7!}{3!4!}$ **(v)** $\dfrac{10!}{4!6!}$
   b The answers are all integers. Can you explain this?
   c Each expression in **a** is the expansion of a function of the form ^nC_r. Express each expansion in this form.
   d Explain why $^nC_r = {}^nC_{n-r}$.

6 a (i) In the card game *Brag*, each player is given 3 cards. Assuming the pack is made up of 52 different cards, how many different hands are possible?
   (ii) In the game *Bridge*, each player is given 13 cards. How many different hands are possible? Give your answer to three significant figures.
   (iii) In the game *Solitaire*, each player is given 52 cards. How many different hands are possible?

b In the national lottery there are 49 numbers from which to pick a set of 6 numbers. How many different sets of 6 numbers can be picked?

c There were 12 people at a meeting. I sat between two. What is the probability that you can pick the two people out correctly from a list of the 12 names?

Another spreadsheet?

	A	B	C	D	E	F	G	H
1		0	1	2	3	4	5	6
2	0	1						
3	1	1	1					
4	2	1	2	1				
5	3	1	3	3	1			
6	4	1	4	6	4	1		
7	5	1	5	10	10	5	1	
8	6	1	6	15	20	15	6	1

1 • Enter 0 in B1.
  • Enter '= B1 + 1' in C1.
  • Fill right.
2 • Enter 0 in A2.
  • Enter '= A2 + 1' in A3.
  • Fill down.
3 • Enter '=FACT($A2)/(FACT(B$1)*FACT($A2-B$1))' in B2.
  • Fill down.
  • Highlight B2 to I16 and fill right.
  • Error messages appear in some upper cells: highlight and delete.

Note
• FACT(n) is the function $n!$.
• Steps 1 and 2 create row and column references.
• Step 3 puts ^nC_r into the cell with reference (n, r).
• The formula may be more recognisable as
$$\frac{\text{FACT}(n)}{\text{FACT}(r).\text{FACT}(n-r)}.$$

This is just Pascal's triangle again!
Remember the rule used to build the triangle is $(n, r-1) + (n, r) = (n+1, r)$.
This can now be re-expressed as: $^nC_{r-1} + {^nC_r} = {^{n+1}C_r}$
^nC_r notation can get quite cumbersome and a more compact notation has been devised, namely $\binom{n}{r}$. In this notation the rule becomes: $\binom{n}{r-1} + \binom{n}{r} = \binom{n+1}{r}$

Proof

$$\binom{n}{r-1} + \binom{n}{r} = \frac{n!}{(r-1)!(n-(r-1))!} + \frac{n!}{r!(n-r)!} \qquad \text{by definition}$$

$$= \frac{n!}{(r-1)!(n-r+1)!} + \frac{n!}{r!(n-r)!}$$

$$= \frac{n!r}{r!(n-r+1)!} + \frac{n!(n-r+1)}{r!(n-r+1)!}$$

$$= \frac{n!r + n!(n-r+1)}{r!(n-r+1)!}$$

$$= \frac{n!(n+1)}{r!(n-r+1)!}$$

$$= \frac{(n+1)!}{r!(n-r+1)!}$$

$$= \binom{n+1}{r} \qquad \text{by definition}$$

> **Note**
> $r \times (r-1)! = r!$ and
> $(n-r+1) \times (n-r)! = (n-r+1)!$

EXERCISE 2B

1 Given that $^{10}C_2 = 45$, $^{10}C_7 = 120$ and $^{10}C_4 = 210$, write down the values of
   a $^{10}C_6$ b $^{10}C_8$ c $^{10}C_3$

2 a From a team of 11 cricketers, 2 are selected to bat first.
     How many different ways can this be done?
   b From a team of 11 cricketers, 9 are selected not to bat first.
     How many different ways can this be done?
   c Complete the statement: $^{11}C_2 =$

3 a A pentagon with diagonals is drawn by joining 5 points on a
     plane, 2 at a time.
     (i) How many ways can 2 points be selected from 5?
     (ii) How many of these joins will be sides of the pentagon?
     (iii) How many of these joins will be diagonals?

   b Prove that an n-sided polygon will have $\binom{n}{2} - n$ diagonals.

   c Counting both sides and diagonals, a polygon is made up of 15 line segments.
     By solving the equation $\binom{n}{2} = 15$, identify the type of polygon.
     (Don't use trial and error.)

4 Solve the following equations.
   a $\binom{n}{2} = 6$ b $\binom{n}{2} = 45$ c $\binom{n}{2} = 28$ d $\binom{n}{2} = 120$

e $\dbinom{2n}{2} = 15$ **f** $\dbinom{2n}{2} = 45$ **g** $\dbinom{2n}{2} = 66$ **h** $\dbinom{2n}{2} = 276$

5 Find a value of n which satisfies each of the following equations.

  a $\dbinom{n}{3} = 4$ **b** $\dbinom{n}{3} = 10$ **c** $\dbinom{n}{3} = 35$ **d** $\dbinom{n}{3} = 120$

6 Make use of the fact that $\dbinom{n}{n-r} = \dbinom{n}{r}$ to help you solve:

  a $\dbinom{n}{n-2} = 15$ **b** $\dbinom{n}{n-2} = 55$ **c** $\dbinom{n}{n-3} = 84$

7 Using the identity $\dbinom{n}{r-1} + \dbinom{n}{r} = \dbinom{n+1}{r}$, find a value of n which satisfies each of the following equations.

  a $\dbinom{n}{1} + \dbinom{n}{2} = 28$ **b** $\dbinom{n+1}{1} + \dbinom{n+1}{2} = 66$

  c $\dbinom{4}{n-1} + \dbinom{4}{n} = 5$ **d** $\dbinom{n+1}{2} - \dbinom{n}{1} = 36$ [careful]

The binomial theorem

At the beginning of the chapter we saw how row n of Pascal's triangle gave us the coefficients of expansions of the form $(x + y)^n$, for example:
$$(x + y)^4 = 1x^4 + 4x^3y + 6x^2y^2 + 4xy^3 + 1y^4$$
In the $\dbinom{n}{r}$ notation, this could have been written as
$$\dbinom{4}{0}x^4 + \dbinom{4}{1}x^3y + \dbinom{4}{2}x^2y^2 + \dbinom{4}{3}xy^3 + \dbinom{4}{4}y^4$$
When this is generalised for any whole number n, we get
$$(x + y)^n = \dbinom{n}{0}x^n + \dbinom{n}{1}x^{n-1}y + \dbinom{n}{2}x^{n-2}y^2 + \cdots + \dbinom{n}{r}x^{n-r}y^r + \cdots + \dbinom{n}{n}y^n$$
This expansion is known as the *binomial theorem*. It is introduced now because it is a very useful result, but the proof of the theorem for positive integers will not appear until Chapter 2.

Historical note

The binomial theorem, where n is a positive integer, was known to the Chinese in the fourteenth century. It was developed by Newton in the seventeenth century to include any rational number.

The proof, where n takes any value, was completed by a Norwegian mathematician, Neils Abel, in the nineteenth century.

Using a special notation, *sigma notation*, the theorem can be quoted in a very compact form

$$(x + y)^n = \sum_{r=0}^{n} \binom{n}{r} x^{n-r} y^r$$

Σ stands for *sum* and is pronounced *sigma*. It acts like an instruction set:
1. create terms using the formula given to the right of Σ by replacing r with each of the integers from 0 to n in turn;
2. add all these terms together.

Example 1   Expand $(1 + x)^5$ using the binomial theorem.

$$\binom{5}{0}1^5x^0 + \binom{5}{1}1^4x^1 + \binom{5}{2}1^3x^2 + \binom{5}{3}1^2x^3 + \binom{5}{4}1^1x^4 + \binom{5}{5}1^0x^5$$

Using the ^nC_r button on a calculator, or otherwise, we get
$$1 + 5x + 10x^2 + 10x^3 + 5x^4 + x^5$$

Example 2   Expand $(1 - 3p)^3$ using the binomial theorem.

$$\binom{3}{0}1^3(-3p)^0 + \binom{3}{1}1^2(-3p)^1 + \binom{3}{2}1^1(-3p)^2 + \binom{3}{3}1^0(-3p)^3$$
$$= 1 + 3(-3p) + 3(9p^2) + (-27p^3)$$
$$= 1 - 9p + 27p^2 - 27p^3$$

EXERCISE 3A

1 Use the binomial theorem to expand the following.
   a $(a + b)^5$ **b** $(1 + 2x)^3$ **c** $(2 + 3b)^4$ **d** $(3a + 2b)^3$
   e $(a - b)^4$ **f** $(1 - p)^3$ **g** $(3 - x)^4$ **h** $(2a - 3b)^3$

2 **a** Expand the following expressing your answer as positive powers of x.

   (i) $\left(x + \dfrac{1}{x}\right)^3$ **(ii)** $\left(x + \dfrac{1}{x}\right)^4$ **(iii)** $\left(x - \dfrac{1}{x}\right)^5$ **(iv)** $\left(x - \dfrac{1}{x}\right)^6$

   b Which of these expressions produced a term independent of x?

3 Work out
   a the third term in the expansion of $(x + y)^{12}$, i.e. the term containing x^{10}
   b the fourth term in the expansion of $(3 + a)^8$
   c the seventh term in the expansion of $(2x + 3y)^9$
   d the second term in the expansion of $(2x + 5)^7$
   e the eighth term in the expansion of $(x - y)^9$
   f the fifth term in the expansion of $(3x - 4y)^5$

4 Calculate the term
   a containing x^4 in the expansion of $(x + y)^8$
   b containing a^3 in the expansion of $(3 + 2a)^5$
   c whose coefficient is 64 in the expansion of $(2 + x)^6$

   d containing x^3 in the expansion of $(x - 7)^5$

   e containing a^4 term in the expansion of $(1 - 3a)^6$

   f independent of x in the expansion of $\left(x + \dfrac{1}{x} \right)^4$

5 a Expand $(1 + x + y)^3$ by first expressing it as $[(1 + x) + y]^3$.

   b In a similar fashion expand **(i)** $(2 + a + 2b)^3$ **(ii)** $(1 - x + y)^5$ **(iii)** $(1 + x - y)^4$

6 By considering $(1 + x)^n$, prove that $2^n = {}^nC_0 + {}^nC_1 + {}^nC_2 + \cdots + {}^nC_n$.

7 Every quadratic expression can be written in the form $a(x + b)^2 + c$ by a process known as *completing the square*.

By considering a similar process write $x^3 + 6x^2 + 10x + 4$ in the form $(x + a)^3 + bx + c$

8 *Remember, Pascal used the binomial theorem for probability theory. In what way?*
If p is the probability of being stopped at any one set of traffic lights and q is the probability of not being stopped, then the terms of the expansion $(p + q)^3$ provide formulae for the probability of being stopped by 3, 2, 1, 0 sets of lights.

$$(p + q)^3 = p^3 + 3p^2q + 3pq^2 + q^3$$

$$P(3) = p^3, \quad P(2) = 3p^2q, \quad P(1) = 3pq^2, \quad P(0) = q^3$$

   a If $p = 0.8$ and $q = 0.2$ calculate the probability of being stopped at

     (i) two sets of lights   **(ii)** all three sets of lights

   b By expanding $(p + q)^4$, find the probability of being stopped at 2 out of 4 sets of lights.

9 The probability that it will rain on any day is 0.4, and that it won't rain is 0.6.
By considering the expansion $(p + q)^7$, calculate the probability that

   a it will rain twice in a 7 day week.

   b it won't rain in the week.

Harder examples

Example 1   What is the coefficient of x^5 in the expansion of $(1 + x)^4(1 - 2x)^3$?

Terms in x^5 are obtained by multiplying certain terms together, namely:
the term containing x^2 in the first expansion with the term containing x^3 in the second,
the term containing x^3 in the first expansion with the term containing x^2 in the second,
the term containing x^4 in the first expansion with the term containing x in the second.

$$\binom{4}{2}x^2\binom{3}{0}x^3 + \binom{4}{1}x^3\binom{3}{1}x^2 + \binom{4}{0}x^4\binom{3}{2}x$$

$$= 6x^2.x^3 + 4x^3.3x^2 + x^4.3x$$

$$= 21x^5$$

The required coefficient is therefore 21.

Example 2   Expand $(x + 1)^2(1 + 2x + x^2)^3$.

Consider each set of brackets separately:

$$(x + 1)^2 = \binom{2}{0}x^2 + \binom{2}{1}x + \binom{2}{2}1 = x^2 + 2x + 1$$

$$(1 + 2x + x^2)^3 = ((1 + 2x) + x^2)^3$$

$$= \binom{3}{0}(1 + 2x)^3 + \binom{3}{1}(1 + 2x)^2(x^2) + \binom{3}{2}(1 + 2x)(x^2)^2 + \binom{3}{3}(x^2)^3$$

$$= (1 + 6x + 12x^2 + 8x^3) + 3(1 + 4x + 4x^2)x^2 + 3(1 + 2x)x^4 + x^6$$

$$= 1 + 6x + 15x^2 + 20x^3 + 15x^4 + 6x^5 + x^6$$

$$(x + 1)^2(1 + 2x + x^2)^3 = (x^2 + 2x + 1)(1 + 6x + 15x^2 + 20x^3 + 15x^4 + 6x^5 + x^6)$$

The expansion can be made clearer by a tabular layout.

multiplying by x^2:	$x^2 + \ 6x^3 + 15x^4 + 20x^5 + \ 15x^6 + 6x^7 + x^8$
multiplying by $2x$:	$2x + 12x^2 + 30x^3 + 40x^4 + 30x^5 + 12x^6 + 2x^7$
multiplying by 1:	$1 + 6x + 15x^2 + 20x^3 + 15x^4 + \ 6x^5 + \ x^6$
Total:	$1 + 8x + 28x^2 + 56x^3 + 70x^4 + 56x^5 + 28x^6 + 8x^7 + x^8$

Therefore
$$(x + 1)^2(1 + 2x + x^2)^3 = \ 1 + 8x + 28x^2 + 56x^3 + 70x^4 + 56x^5 + 28x^6 + 8x^7 + x^8$$

EXERCISE 3B

1 What is the coefficient of
   a x^4 in the expansion $(1 + x)^2(1 + 2x)^3$
   b x^5 in the expansion $(1 - x)^3(2 + x)^4$
   c x^7 in the expansion $(1 + 2x)^4(1 - 2x)^6$?

2 Find the coefficients of x^3 and x^5 in the expansion of $(1 + x + x^2)^5$.

3 What are the terms in x^3 and x^{10} in $(1 + x)^5(1 - x + x^2)^4$?

4 Expand
   a $(1 + x)^2(1 + x + x^2)^3$
   b $(1 - 3x)^3(1 + 2x + x^2)^2$
   c $(3 + 2x)^2(1 - x + x^2)^4$
   d $(1 - 2x)^4(1 + 2x + 4x^2)^3$

5 **a** Expand $\left(x + \dfrac{1}{x}\right)^2\left(x - \dfrac{1}{x}\right)^3$.
   b Expand $\left(x + \dfrac{1}{x}\right)^2\left(x - \dfrac{1}{x}\right)^4$.
   c Under what circumstances do you get terms independent of x?
   d Find the term independent of x in the expansion of $\left(x + \dfrac{1}{x}\right)^6\left(x - \dfrac{1}{x}\right)^8$.

6 Find the terms in a^5 and a^6 in $\left(3a^2 - \dfrac{1}{a}\right)^6\left(a + \dfrac{1}{a}\right)^4$.

7 Find the term independent of a in $\left(\dfrac{3}{2}a^2 - \dfrac{1}{3a}\right)^9$.

8 **a** Use the binomial theorem to help you write down expressions for the coefficients of x^r and x^{r+1} in the expansion of $(3x + 2)^{19}$.
   b Find the value of r if these coefficients are equal.

9 If $x = \dfrac{1}{4}$, find the ratio of the 8th and 7th terms in the expansion of $(1 + 2x)^{15}$.

10 Which are the greatest terms in the following expansions?
   a $(1 + 3x)^{18}$ when $x = \dfrac{1}{4}$
   b $\left(1 + \dfrac{1}{2}x\right)^{12}$ when $x = \dfrac{1}{2}$
   c $(4 + x)^8$ when $x = 3$
   d $(x + y)^n$ when $n = 14$, $x = 2$, $y = \dfrac{1}{2}$

11 Find the numerically greatest terms in the following expansions.
   a $(2 - x)^{12}$ when $x = \dfrac{2}{3}$
   b $(3a + 2b)^n$ when $n = 16$, $a = 1$, $b = \dfrac{1}{2}$

12 Find which terms have the greatest coefficients in:
   a $(1 + x)^{10}$
   b $(2 + x)^{11}$
   c $(1 + x)^{2n+1}$

13 In the following expansion, show that there are two greatest terms and find their values.
   $(a + x)^n$ when $n = 9$, $a = \dfrac{1}{2}$, $x = \dfrac{1}{3}$

14 Find the coefficients of x^2 and x^3 in the expansion of $(2 + 2x + x^2)^n$.

15 Expand $(1 - x + x^2)^n$ in ascending powers of x as far as the term in x^3.

Approximation

For $-1 < a < 1$, as $n \to \infty$ then $a^n \to 0$.
Using this fact allows us to make useful approximations.

Example 1   Calculate 1.02^3 correct to three decimal places.

$$1.02^3 = (1 + 0.02)^3 = 1 + 3 \times 0.02 + 3 \times 0.02^2 + 0.02^3$$
$$= 1 + 0.06 + 0.0012 + 0.000\,008$$
$$= 1.061 \text{ to 3 d.p.}$$

Note that the term 0.02^3 does not contribute to the rounded result.

Example 2   Calculate 0.9^7 correct to two decimal places.

$$0.9^7 = (1 - 0.1)^7 = 1 - 7 \times 0.1 + 21 \times 0.1^2 - 35 \times 0.1^3 + 35 \times 0.1^4 \ldots$$
$$= 1 - 0.7 + 0.21 - 0.035 + 0.0035 \ldots$$
$$= 0.48 \text{ to 2 d.p.}$$

Note that the terms in 0.1^4 and higher do not contribute to the rounded result.

EXERCISE 4

1 Calculate the following correct to three significant figures.

  a 1.01^5 **b** 1.04^6 **c** 0.94^7 **d** 12.01^5

  e e^8 **f** π^7 **g** $\sin^6 30°$ **h** 199^4

2

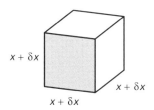

When metal of length x is heated, it expands by the amount δx, where δx is so small that $(\delta x)^2$ and higher powers can be ignored.

Write down an expression for

  a the increase in area when a square of metal of side x expands;

  b the increase in volume when a cube of metal of side x expands.

3 Given that a is so small that terms in a^3 are negligible, work out approximate expansions for:

  a $(1 + a)^6$ **b** $(2a - 1)^7$ **c** $(a^2 + 2)^8$

4 Given that x is so small that x^3 and higher powers can be neglected, show that

$$\left(1 - \frac{3}{2}x\right)^5 (2 + 3x)^6 \approx 64 + 96x - 720x^2$$

5 Assuming that $f'(x) \approx \dfrac{f(x + h) - f(x)}{h}$ where h is so small that h^2 and higher powers can be ignored, use the binomial theorem to find $f'(x)$ when $f(x) = x^n$.

6 a The value, in pounds (£), of a car, n years old, can be computed from the formula

$$V_n = 0.9^n \times 10\,000$$

    Use the binomial theorem to come up with a simplified formula which will be useful when only two significant figures are required.

  b In a similar manner, simplify the following formula where V_n is the value of an investment after n years.

$$V_n = 1.05^n \times 5000.$$

CHAPTER 1.1 REVIEW

1 Write out Pascal's triangle to the seventh row.

2 Consider the expansion of $(x + y)^{20}$.

   a The term in x^5 is of the form $\begin{pmatrix} p \\ q \end{pmatrix} x^5 y^r$. What are the values of p, q and r?

   b For what values of s and t is $\begin{pmatrix} 20 \\ 14 \end{pmatrix}$ the coefficient of $x^s y^t$?

   c Evaluate (i) $\begin{pmatrix} 20 \\ 14 \end{pmatrix}$ (ii) $^{20}C_{11}$

3 a For what value of $n \neq 3$ is $\begin{pmatrix} 8 \\ n \end{pmatrix} = \begin{pmatrix} 8 \\ 3 \end{pmatrix}$?

   b For what values of p and q is $\begin{pmatrix} 7 \\ 3 \end{pmatrix} + \begin{pmatrix} 7 \\ 4 \end{pmatrix} = \begin{pmatrix} p \\ q \end{pmatrix}$?

4 Use the binomial theorem to expand
   a $(x - 4)^5$
   b $(2y - 3)^3$

5 a Evaluate the term in x^7 in the expansion of $(1 + x)^{10}$.

   b Evaluate the term in y^3 in the expansion of $\left(y - \dfrac{5}{y} \right)^7$

6 Show that, in the expansion of $\left(x - \dfrac{2}{x} \right)^{10}$, the term which is independent of x has the value -8064.

7 a Evaluate $^{11}C_0 + {}^{11}C_1 + {}^{11}C_2 + \cdots + {}^{11}C_{10} + {}^{11}C_{11}$ by considering the expansion of $(1 + x)^{11}$ when $x = 1$.
   b Evaluate
      (i) $^{11}C_0 + {}^{11}C_2 + {}^{11}C_4 + \cdots + {}^{11}C_8 + {}^{11}C_{10}$
      (ii) $^{11}C_1 + {}^{11}C_3 + {}^{11}C_5 + \cdots + {}^{11}C_9 + {}^{11}C_{11}$
      by considering the expansion of $(1 + x)^{11}$ when $x = -1$.

8 Prove that $\begin{pmatrix} n \\ r \end{pmatrix} + 2 \begin{pmatrix} n \\ r + 1 \end{pmatrix} + \begin{pmatrix} n \\ r + 2 \end{pmatrix} = \begin{pmatrix} n + 2 \\ r + 2 \end{pmatrix}$.

CHAPTER 1.1 SUMMARY

1 $n! = n \times (n-1) \times (n-2) \times \ldots \times 3 \times 2 \times 1$

2 $\dbinom{n}{r} = {}^{n}C_r = \dfrac{n!}{r!(n-r)!}$

3 $\dbinom{n}{r} = \dbinom{n}{n-r}$

4 $\dbinom{n}{r-1} + \dbinom{n}{r} = \dbinom{n+1}{r}$

5 Pascal's triangle

	0	1	2	3	4	5	6	7
0	1							
1	1	1						
2	1	2	1					
3	1	3	3	1				
4	1	4	6	4	1			
5	1	5	10	10	5	1		
6	1	6	15	20	15	6	1	
7	1	7	21	35	35	21	7	1

- Each entry can be computed using ${}^{\text{row}}C_{\text{column}}$ (see summary point 2).
- Each row is symmetrical (see summary point 3).
- Each entry is the sum of the entry above it and the entry above it and to the left (see summary point 4).

6 The binomial theorem states:

$$(x+y)^n = \sum_{r=0}^{n} \binom{n}{r} x^{n-r} y^r$$

which when expanded becomes

$$(x+y)^n = \binom{n}{0}x^n + \binom{n}{1}x^{n-1}y + \binom{n}{2}x^{n-2}y^2 + \cdots + \binom{n}{r}x^{n-r}y^r + \cdots + \binom{n}{n}y^n$$

7 The general term in the binomial expansion is $\dbinom{n}{r}x^{n-r}y^r$.

This can be used to find particular terms with particular properties.

1.2 Partial Fractions

Proper and improper rational functions; algebraic division

A rational function is one expressed in fractional form whose numerator and denominator are polynomials.

A rational function is termed *proper* when the degree of the numerator is less than the degree of the denominator.

It is termed *improper* otherwise.

For example $\dfrac{x + 1}{x^2 + 2}$, $\dfrac{x}{(x + 1)(x + 2)}$ and $\dfrac{2}{x^2 + 3x + 1}$ are all *proper* rational functions

whereas $\dfrac{x^3 + 1}{x^2 + 2}$, $\dfrac{x}{x + 2}$ and $\dfrac{x^2}{(x + 1)(x + 2)}$ are all *improper* rational functions.

Improper rational functions can be simplified by algebraic division.

Example $\dfrac{x^3 + 4x^2 - x + 2}{x^2 + x} = (x^3 + 4x^2 - x + 2) \div (x^2 + x)$

$$
\begin{array}{r}
x \\
x^2 + x \overline{\smash{)}\ x^3 + 4x^2 - x + 2} \\
\underline{x^3 + x^2} \\
3x^2 - x + 2
\end{array}
$$

By what do we multiply x^2 to make x^3?

Answer: x

Multiply divisor by x.

Subtract result from dividend to find remainder.

$$
\begin{array}{r}
x + 3 \\
x^2 + x \overline{\smash{)}\ x^3 + 4x^2 - x + 2} \\
\underline{x^3 + x^2} \\
3x^2 - x + 2 \\
\underline{3x^2 + 3x} \\
-4x + 2
\end{array}
$$

By what do we multiply x^2 to make $3x^2$?

Answer: 3

Multiply divisor by 3

Subtract to find remainder.

The remainder is now a lower order than the divisor, so

$$\frac{x^3 + 4x^2 - x + 2}{x^2 + x} = x + 3 + \frac{-4x + 2}{x^2 + x}$$

The fractional part is now a proper rational function

EXERCISE 1

Simplify the following rational functions by algebraic division.

1 $\dfrac{x^2 + 3x + 5}{x + 2}$ **2** $\dfrac{x^2 - 2x + 4}{x + 3}$ **3** $\dfrac{x^2 + 3x - 5}{x - 2}$ **4** $\dfrac{3x^2 - 5x + 1}{x - 4}$

5 $\dfrac{3x^2 - 4x + 5}{x^2 + x + 1}$ **6** $\dfrac{x^2 - x + 1}{x^2 + x - 2}$ **7** $\dfrac{x^2}{x^2 - x + 2}$ **8** $\dfrac{x^3 + 3x^2 + 4x - 5}{x^2 + 1}$

9 $\dfrac{x^3 + 1}{x + 2}$ [Hint: express the numerator as $x^3 + 0x^2 + 0x + 1$.]   **10** $\dfrac{3x^3 - 2x + 4}{x - 4}$

11 $\dfrac{x^2 + 3}{x^2 - 4}$ **12** $\dfrac{x^4 + 3x^3 + 2x^2 - 3}{x^2 + 2x}$ **13** $\dfrac{x^5 + 1}{x^3 - x + 1}$ **14** $\dfrac{3x^3 + 7x - 1}{x^2 + 3}$

General forms

In the general form of a proper rational function the numerator has an order 1 less than the denominator. For example, if the denominator is linear then the general form will be

$$\frac{a}{bx + c} \quad a \text{ and } c \text{ take any integral value; } b \neq 0;\ bx + c \neq 0$$

if the denominator is quadratic then the general form will be

$$\frac{ax + b}{cx^2 + dx + e} \quad a, b, d \text{ and } e \text{ take any integral value; } c \neq 0;\ cx^2 + dx + e \neq 0$$

Distinct linear factors in the denominator

When we encounter a rational function whose denominator is the product of two functions it is often of great help to decompose the function into the sum of two simpler rational functions. These simpler functions are referred to as *partial fractions*. Their denominators will be the factors of the denominator of the original function. The simplest case to study is when the factors of the denominator are linear and distinct.

Example   Decompose $\dfrac{4x + 1}{(x + 1)(x - 2)}$ into partial fractions.

$$\frac{4x + 1}{(x + 1)(x - 2)} = \frac{A}{(x + 1)} + \frac{B}{(x - 2)} \qquad \textit{general forms}$$

Multiply throughout by $(x + 1)(x - 2)$.

$$4x + 1 = A(x - 2) + B(x + 1)$$

We are looking for values of A and B which make this statement true for all values of x. We can select any value of x we please. By inspection we see that
- if $x = 2$ then the equation becomes $9 = 3B$, thus B = 3;
- if $x = -1$ then the equation becomes $-3 = -3A$, thus $A = 1$.

$$\frac{4x + 1}{(x + 1)(x - 2)} = \frac{1}{(x + 1)} + \frac{3}{(x - 2)}$$

EXERCISE 2

Resolve each of the following rational functions into its partial fractions.

1 $\dfrac{2x}{(x-1)(x+1)}$

2 $\dfrac{10x}{(x-2)(x+3)}$

3 $\dfrac{4x}{(x+1)(x+5)}$

4 $\dfrac{20}{(x-3)(x+2)}$

5 $\dfrac{3}{(x-1)(x+2)}$

6 $\dfrac{5}{(x+2)(x+3)}$

7 $\dfrac{5x-14}{(x-2)(x-3)}$

8 $\dfrac{3x-2}{(x+2)(x-2)}$

9 $\dfrac{15-2x}{(2x-1)(x+3)}$

10 $\dfrac{7x+5}{(x-1)(x+3)}$

11 $\dfrac{-x-7}{(x-3)(x+1)}$

12 $\dfrac{6x-15}{x(x+3)}$

13 $\dfrac{3x+4}{x(x+2)}$

14 $\dfrac{-3x-12}{x(x+6)}$

15 $\dfrac{4-6x}{x(x+4)}$

16 $\dfrac{7x-11}{(5-x)(x+1)}$

17 $\dfrac{11x+6}{(3x+1)(2x+3)}$

18 $\dfrac{2x+7}{(2x-3)(4x-1)}$

19 $\dfrac{5x-3}{x^2+x-30}$

20 $\dfrac{5x-11}{x^2-4x+3}$

21 $\dfrac{4x+5}{x^2+3x+2}$

22 $\dfrac{4x-10}{x^2-3x}$

23 $\dfrac{6x-1}{4x^2-1}$

24 $\dfrac{4x+1}{2x^2+3x+1}$

25 $\dfrac{2x^2-2x-6}{(x+1)(x+2)(x-1)}$

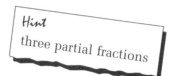

Hint
three partial fractions

26 $\dfrac{6x^2-10x+2}{x^3-3x^2+2x}$

A repeated linear factor in the denominator

When a factor of the form $(ax+b)^2$ appears in the denominator, we would expect
the general form of the partial fraction to be $\dfrac{cx+d}{(ax+b)^2}$

i.e. with the numerator of order 1 given that the denominator is order 2. However,
notice that

$$\frac{cx+d}{(ax+b)^2} = \frac{\dfrac{c}{a}(ax+b) - \dfrac{cb}{a} + d}{(ax+b)^2}$$ Check this by getting rid of the brackets
in the numerator.

$$= \frac{\dfrac{c}{a}(ax+b) + \left(d - \dfrac{cb}{a}\right)}{(ax+b)^2}$$

$$= \frac{\dfrac{c}{a}}{(ax+b)} + \frac{\left(d - \dfrac{cb}{a}\right)}{(ax+b)^2}$$

Since a, b, c and d are constants, we see that the general form in this case has constant numerators, i.e.

$$\frac{cx + d}{(ax + b)^2} = \frac{A}{(ax + b)} + \frac{B}{(ax + b)^2} \quad \text{for some constants } A \text{ and } B$$

Example   Reduce $\dfrac{x^2 - 7x + 9}{(x + 2)(x - 1)^2}$ to its partial fractions.

$$\frac{x^2 - 7x + 9}{(x + 2)(x - 1)^2} = \frac{A}{x + 2} + \frac{B}{x - 1} + \frac{C}{(x - 1)^2}$$

Multiplying throughout by $(x + 2)(x - 1)^2$ gives:

$$x^2 - 7x + 9 = A(x - 1)^2 + B(x + 2)(x - 1) + C(x + 2)$$

- Let $x = -2$: $(-2)^2 - 7 \times (-2) + 9 = A \times (-3)^2$ which gives $A = 3$.
- Let $x = 1$: $(1)^2 - 7 \times (1) + 9 = C \times 3$ which gives $C = 1$.
- Let $x = 0$: $(0)^2 - 7 \times (0) + 9 = A + B \times (-2) + C \times 2$.

(an arbitrary choice)   Substituting $A = 3$ and $C = 1$ gives $B = -2$.

$$\frac{x^2 - 7x + 9}{(x + 2)(x - 1)^2} = \frac{3}{x + 2} - \frac{2}{x - 1} + \frac{1}{(x - 1)^2}$$

EXERCISE 3

Express each of the following in partial fractions.

1 $\dfrac{3x^2 - 11x + 5}{(x - 2)(x - 1)^2}$

2 $\dfrac{6x^2 + x - 7}{(x + 2)^2(x - 3)}$

3 $\dfrac{4x^2 - 8x + 15}{(x - 2)^2(x + 1)}$

4 $\dfrac{x^2 - x - 1}{x^2(x - 1)}$

5 $\dfrac{x^2 + 6x - 3}{x(x - 1)^2}$

6 $\dfrac{4x^2 + 19x + 13}{(x - 1)(x + 2)^2}$

7 $\dfrac{2x + 5}{(1 - x)(x + 2)^2}$

8 $\dfrac{25}{(2 - x)(2x + 1)^2}$

9 $\dfrac{5x + 1}{x(2x + 1)^2}$

10 $\dfrac{x - 2}{x^2(3x - 2)}$

11 $\dfrac{1}{x^2 - 3x^3}$

12 $\dfrac{16}{(x^2 - 2x - 3)(x - 3)}$

13 $\dfrac{-x^2 + 7x - 6}{x^3 - 2x^2}$

14 $\dfrac{3x^2 - 26x + 77}{(2x + 1)(x - 5)^2}$

15 $\dfrac{3x^2 - 14x - 1}{(x^2 + 4x - 5)(x - 1)}$

16 $\dfrac{7x^2 + 1}{x^3 - x^2}$

17 $\dfrac{x^2 - 9x - 2}{(x - 1)(x^2 - 1)}$

18 $\dfrac{2x^2 + 7x + 3}{x^3 + 2x^2 + x}$

An irreducible quadratic factor in the denominator

Consider the function $f(x) = x^2 + 2x + 2$.
Its discriminant is $2^2 - 4 \times 1 \times 2 = -4$.
As this is less than zero, the function has no
real roots and thus the expression $x^2 + 2x + 2$
has no real factors (it is *irreducible*).

If an irreducible expression appears as a factor in
the denominator of a rational function which has
to be decomposed into its partial fractions, then the general form

$$\frac{ax + b}{cx^2 + dx + e}$$

has to be considered.

> **Reminder**
> If $f(x) = ax^2 + bx + c$,
> the discriminant is $b^2 - 4ac$.

Example   Reduce $\dfrac{3x^2 + 2x + 1}{(x + 1)(x^2 + 2x + 2)}$ into its partial fractions.

$$\frac{3x^2 + 2x + 1}{(x + 1)(x^2 + 2x + 2)} = \frac{A}{x + 1} + \frac{Bx + C}{x^2 + 2x + 2}$$

Multiplying throughout by $(x + 1)(x^2 + 2x + 2)$ gives

$$3x^2 + 2x + 1 = A(x^2 + 2x + 2) + (Bx + C)(x + 1)$$

We are looking for values of A, B and C which make this true for any x.
Choose suitable values for x.
- To eliminate $(Bx + C)$, let $x = -1$: $2 = A + 0 \Rightarrow A = 2$
- To eliminate Bx, let $x = 0$: $1 = 2A + C \Rightarrow C = -3$
- Let $x = 1$ (an arbitrary choice): $6 = 5A + 2B + 2C \Rightarrow B = 1$

Thus: $\dfrac{3x^2 + 2x + 1}{(x + 1)(x^2 + 2x + 2)} = \dfrac{2}{x + 1} + \dfrac{x - 3}{x^2 + 2x + 2}$

EXERCISE 4

Express each of the following in partial fractions.

1 $\dfrac{3}{(x + 1)(x^2 - x + 1)}$

2 $\dfrac{8}{(1 - x)(x^2 + 2x + 5)}$

3 $\dfrac{16x^2}{(x - 3)(2x^2 - x + 1)}$

4 $\dfrac{2x^2 + 3x + 1}{(x - 1)(x^2 + x + 1)}$

5 $\dfrac{2x^2 - 11}{(x - 3)(x^2 - x + 1)}$

6 $\dfrac{3x - 2}{x^3 + 2x}$

7 $\dfrac{x^2 + 2x + 9}{(x^2 + 3)(x - 1)}$

8 $\dfrac{4x^2 + 5x + 13}{(x^2 + x + 3)(x + 1)}$

9 $\dfrac{4x^2 - 4x + 1}{(x^2 - x + 1)(x - 2)}$

10 $\dfrac{x^2 + 3x + 3}{(2x + 1)(x^2 + x + 2)}$

11 $\dfrac{x^2 + x}{x^3 - 2x^2 + 2x - 1}$

12 $\dfrac{8x^2 - 5x + 6}{x^3 + x^2 + 4}$

13 $\dfrac{x^2 - 2x + 2}{x^3 - 1}$

14 $\dfrac{x^2 - 10x - 8}{x^3 - 8}$

15 $\dfrac{4x^2 - 3x + 2}{x^3 - 1}$

Before attempting to resolve a proper rational function into its partial fractions the denominator should be carefully examined to see what general forms are to be used.

EXERCISE 5

Resolve each proper rational function into its partial fractions.

1 $\dfrac{x}{(1-x)(2+x)}$

2 $\dfrac{2x-1}{(2x+1)(x-3)}$

3 $\dfrac{3x}{(x-2)(x+1)}$

4 $\dfrac{2}{(x-1)^2(x+1)}$

5 $\dfrac{3x^2-4}{x(x^2+1)}$

6 $\dfrac{3}{x(x-2)^2}$

7 $\dfrac{1}{x(x^2+4)}$

8 $\dfrac{4x-3}{x^3(x+1)}$

9 $\dfrac{5x-3}{(x+2)(x-3)^2}$

10 $\dfrac{3x^2+2x}{(x+2)(x^2+3)}$

11 $\dfrac{3}{1-x^3}$

12 $\dfrac{x^2+1}{x(x^2-1)}$

13 $\dfrac{2x-1}{(x-2)(x+1)(x+3)}$

14 $\dfrac{4x-1}{x^2(x^2-4)}$

15 $\dfrac{1}{x^2-2}$

16 $\dfrac{x^2}{(x-3)^3}$

17 $\dfrac{(x+13)^2}{(x-3)^2(x+5)}$

18 $\dfrac{1-2x}{x^3+1}$

19 $\dfrac{x}{x^4-16}$

20 $\dfrac{2x-1}{(x-3)^2(x+5)}$

21 $\dfrac{3x}{(x+1)(3-x^2)}$

22 $\dfrac{2x^2-5x}{(x^2-1)(x^2-4)}$

23 $\dfrac{1}{x^3(1-2x)}$

24 $\dfrac{2x-7}{(x^2+4)(x-1)^2}$

25 $\dfrac{1}{x(x^2-1)^2}$

26 $\dfrac{1}{x(x^2+4)^2}$

Handling improper rational functions

As demonstrated at the beginning of this section on partial fractions, an improper rational function can be reduced, by algebraic division, to the sum of a polynomial function and a *proper* rational function. This proper rational function can then be resolved into its partial fractions.

Example   Resolve $\dfrac{x^3 + 4x^2 - x + 2}{x^2 + x}$ into a polynomial function plus partial fractions.

$$
\begin{array}{r}
x + 3 \\
x^2 + x\ \overline{\smash{\big)}\ x^3 + 4x^2 - x + 2} \\
\underline{x^3 + x^2} \\
3x^2 - x + 2 \\
\underline{3x^2 + 3x} \\
-4x + 2
\end{array}
$$

The division gives us $\dfrac{x^3 + 4x^2 - x + 2}{x^2 + x} = x + 3 + \dfrac{-4x + 2}{x^2 + x}$

Resolving the rational portion into partial fractions $\dfrac{-4x + 2}{x^2 + x} = \dfrac{-4x + 2}{x(x + 1)} = \dfrac{A}{x} + \dfrac{B}{x + 1}$

$$\Rightarrow -4x + 2 = A(x + 1) + Bx$$
$$x = 0 \Rightarrow A = 2$$
$$x = -1 \Rightarrow B = -6$$

Thus $\dfrac{x^3 + 4x^2 - x + 2}{x^2 + x} = x + 3 + \dfrac{2}{x} + \dfrac{-6}{x + 1} = x + 3 + \dfrac{2}{x} - \dfrac{6}{x + 1}$

EXERCISE 6

1   Resolve each of the improper rational functions into a polynomial function plus partial fractions.

a $\dfrac{x^2 - x + 6}{x^2 + x - 2}$

b $\dfrac{x^3 - x^2 - 5x - 7}{x^2 - 2x - 3}$

c $\dfrac{x^3 - 5x^2 + 11x - 12}{x^2 - 5x + 6}$

d $\dfrac{2x^2 - 7}{x^2 - 4}$

e $\dfrac{x^3 - 3x}{x^2 - x - 2}$

f $\dfrac{x^2}{(x - 1)^2}$

g $\dfrac{x^3 + 2}{x(x^2 - 3)}$

h $\dfrac{x^4 + 1}{x^3 + 2x}$

i $\dfrac{3x^4 + 4x^3 + 6x^2 + 2x - 1}{x^3 + x^2}$

j $\dfrac{(x + 2)(x - 2)}{(x + 1)(x - 1)}$

k $\dfrac{(x + 3)(x - 1)}{(x + 2)(x + 1)}$

l $\dfrac{(x + 1)(x - 2)(x + 3)}{(x - 1)(x - 3)}$

2   Assuming $\dfrac{x^2}{(x + a)(x + b)}$ can be expressed in the form $A + \dfrac{B}{(x + a)} + \dfrac{C}{(x + b)}$ find expressions for A, B and C in terms of a and b.

CHAPTER 1.2 REVIEW

1 Resolve the following into partial fractions.

   a $\dfrac{x + 14}{(x - 4)(x + 2)}$

   b $\dfrac{1 + x - 3x^2}{(x - 2)(x + 1)^2}$

   c $\dfrac{5x^2 + 6x + 7}{(x - 1)(x^2 + 2x + 3)}$

2 Factorise the denominator and then resolve the following rational functions into partial fractions.

   a $\dfrac{x + 3}{x^3 - x}$

   b $\dfrac{x^2 - 3x + 3}{x^2 - x^3}$

   c $\dfrac{1 - 2x - x^2}{x^3 + x^2 + x}$

3 Express each improper rational function as the sum of a polynomial function and partial fractions.

   a $\dfrac{x^2 + 2}{(x - 1)(x + 2)}$

   b $\dfrac{x^4 + 2x^2 - 2x + 1}{x^3 + x}$

CHAPTER 1.2 SUMMARY

1   A proper rational function can be resolved into partial fractions.
There are three basic cases:

(i)   The factors of the denominator are distinct linear functions.

$$\frac{ax + b}{(cx + d)(ex + f)} = \frac{A}{(cx + d)} + \frac{B}{(ex + f)}$$

(ii) The denominator contains a repeated linear factor

$$\frac{ax^2 + bx + c}{(dx + e)(fx + g)^2} = \frac{A}{(dx + e)} + \frac{B}{(fx + g)} + \frac{C}{(fx + g)^2}$$

(iii) The denominator contains an irreducible quadratic factor.

$$\frac{ax^2 + bx + c}{(dx + e)(fx^2 + gx + h)} = \frac{A}{(dx + e)} + \frac{Bx + C}{(fx^2 + gx + h)}$$

2   In particular examples the values of the upper case constants can be ascertained by suitable selection of convenient values of x.

3   An improper rational function can be reduced to a polynomial and a proper rational function by the process of algebraic division.

2.1 Differential Calculus

Historical note

Newton

Leibniz

Newton invented infinitesimal calculus, devising his method of fluxions about 1665.

Leibniz independently invented calculus, publishing a paper on the subject in 1684. A dispute arose as to who had invented calculus first.

A committee of inquiry set up by the Royal Society in 1712 found in favour of Newton. It is, however, the notation devised by Leibniz which is popularly used.

Fermat

While Newton and Leibniz can share credit for the invention of calculus, Fermat had made critically important discoveries over 10 years before either of them was born. In particular he found the equations of tangents, located stationary points and found the areas under many different curves.

Reminders

Definition

The derivative, or derived function, of $f(x)$, denoted by $f'(x)$, is defined as

$$f'(x) = \lim_{h \to 0}\left(\frac{f(x + h) - f(x)}{h}\right)$$

Geometric interpretation

Given the points $P(a, f(a))$ and $Q(a + h, f(a + h))$

the gradient of PQ, $m_{PQ} = \dfrac{f(a + h) - f(a)}{h}$.

As $h \to 0$, Q approaches P along the curve and the gradient of PQ approaches the gradient of PT, the tangent at P.
Thus the gradient of the tangent at P can be defined as

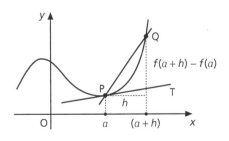

$$\lim_{h \to 0}\left(\frac{f(a + h) - f(a)}{h}\right)$$

which is, of course, the value of the derived function at $x = a$.

Reminder

Leibniz notation

In the definition of $f(x)$, h represents a small change in x and is often referred to as the x-increment and denoted by δx.

$f(x + h) - f(x)$ denotes the corresponding change in y, the y-increment, which is denoted by δy.

Thus we have an alternative notation, Leibniz notation: $f'(x) = \lim\limits_{\delta x \to 0}\left(\dfrac{\delta y}{\delta x}\right) = \dfrac{dy}{dx}$

Strictly speaking, as Q approaches P from the right, and the limit $f'(x)$ exists, we find only the *right* derivative. Similarly if Q approaches P from the left, and the limit $f'(x)$ exists, we find the *left* derivative.

If both limits exist, and are the same, then we may conclude that the function is differentiable.

Differentiation from first principles

Example

Given that $f(x)$ is differentiable, we can use the above definition to prove that, if $f(x) = x^2$, then $f'(5) = 10$, i.e. we can show that the derivative at (5, 25) is 10.

Since we are told that the function is differentiable, we need only consider, say, the right derivative.

$$\frac{f(5 + h) - f(5)}{h} = \frac{(5 + h)^2 - 5^2}{h}$$

$$= \frac{25 + 10h + h^2 - 25}{h}$$

$$= \frac{h(10 + h)}{h}$$

$$= 10 + h$$

$$\lim_{h \to 0}\left(\frac{f(5 + h) - f(5)}{h}\right) = \lim_{h \to 0}(10 + h)$$

$$f'(5) = 10$$

This technique is known as 'finding the derivative from first principles'.

It would be beneficial to find a general rule for the derivative i.e. find the derivative at the point $(x, f(x))$. This 'new' function is called the derived function.

However, much care requires to be exercised. Although polynomial functions are well behaved, other functions such as the tan function have breaks in the graph where the derivative is undefined.

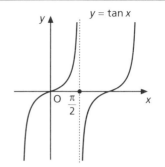

Another example is the function $\sqrt{x^2}$ shown here.

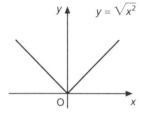

Although the function is continuous, the derivative at O is undefined as the left derivative is negative and the right derivative is positive.

Also there are problems with functions which are only defined on a limited domain. This diagram shows the graph of $y = \sqrt{x}$ which is only defined for $x \geq 0$. Since we cannot calculate the left derivative at $x = 0$ then, at this stage, we shall say that it is differentiable over the interval $x > 0$ (i.e. it is not differentiable at $x = 0$).

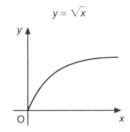

More advanced texts need to be studied to cover the finer points of continuity.

Example

Given that f is differentiable, use the definition of the derivative and the first principles method to prove that, if $f(x) = x^2$, $x \in R$, then $f'(x) = 2x$. (The symbol $\in$ means 'is a member of'.)

$$\frac{f(x + h) - f(x)}{h} = \frac{(x + h)^2 - x^2}{h}$$

$$= \frac{x^2 + 2xh + h^2 - x^2}{h}$$

$$= \frac{h(2x + h)}{h}$$

$$= 2x + h$$

$$\lim_{h \to 0}\left(\frac{f(x + h) - f(x)}{h}\right) = \lim_{h \to 0}(2x + h)$$

$$f'(x) = 2x$$

Limits

Many questions that involve finding the derivative from first principles need some knowledge of limits.

Five rules are given here (without proof).

1 If $f(x) = c$ (a constant), then $\lim_{x \to a} f(x) = c$

2 $\lim_{x \to a} kf(x) = k \times \lim_{x \to a} f(x)$ where k is a constant

3 $\lim_{x \to a} [f(x) \pm g(x)] = \lim_{x \to a} f(x) \pm \lim_{x \to a} f(x)$

4 $\lim_{x \to a} [f(x).g(x)] = \lim_{x \to a} f(x).\lim_{x \to a} g(x)$

5 $\lim_{x \to a} \left[\dfrac{f(x)}{g(x)} \right] = \dfrac{\lim_{x \to a} f(x)}{\lim_{x \to a} g(x)}$ provided $\lim_{x \to a} g(x) \neq 0$.

Two other limits will also occur. Let's see if we can get an idea of what they might be.

a Use your calculator to find some values of $\dfrac{\sin h}{h}$ for smaller and smaller values of h.
You could start with, say, $h = 0.1$.
Remember to set your calculator to radian mode (see the note below).

Can you make a conjecture about $\lim_{h \to 0} \left(\dfrac{\sin h}{h} \right)$?

b In a similar way what can you find out about $\lim_{h \to 0} \left(\dfrac{\cos h - 1}{h} \right)$?

You should have come to the conclusion that

6 $\lim_{h \to 0} \left(\dfrac{\sin h}{h} \right) = 1$

7 $\lim_{h \to 0} \left(\dfrac{\cos h - 1}{h} \right) = 0$

Again we state these results without formal proof.

Radian Reminder

If h is measured in degrees then you would arrive at a different value for $\lim_{h \to 0} \left(\dfrac{\sin h}{h} \right)$. (In fact the limit is approximately 0.001 745 or $\dfrac{2\pi}{360}$ precisely.)
This would then lead to a different set of rules for differentiating all of the trigonometric functions – i.e. $\dfrac{d}{dx}(\sin x°) \neq \cos x°$!
So, whenever calculus is likely to appear, work in radians.

Example   Given that f is differentiable, prove that, if $f(x) = \sin x$, then $f'(x) = \cos x$.

$$\frac{f(x + h) - f(x)}{h} = \frac{\sin(x + h) - \sin x}{h} \qquad \text{by definition}$$

$$= \frac{\sin x \cos h + \cos x \sin h - \sin x}{h} \qquad \text{expanding } \sin(a + b)$$

$$= \frac{\sin x(\cos h - 1) + \cos x \sin h}{h}$$

$$\lim_{h \to 0}\left(\frac{f(x + h) - f(x)}{h}\right) = \lim_{h \to 0}\left[\frac{\sin x(\cos h - 1)}{h} + \frac{\cos x \sin h}{h}\right]$$

$$= \lim_{h \to 0}\left[\frac{\sin x(\cos h - 1)}{h}\right] + \lim_{h \to 0}\left[\frac{\cos x \sin h}{h}\right] \qquad \text{Limit rule 3}$$

$$= \sin x \times \lim_{h \to 0}\left[\frac{\cos h - 1}{h}\right] + \cos x \lim_{h \to 0}\left[\frac{\sin h}{h}\right] \qquad \text{Limit rule 2}$$

$$f'(x) = \sin x \times 0 + \cos x \times 1 \qquad \text{Limit rules 6 and 7}$$

$$= \cos x$$

You can assume that each of the functions in Exercises 1A and 1B is differentiable.

EXERCISE 1A

Find the derivative of each of the following from first principles.

1 $f(x) = 5x$ **2** $f(x) = 12x + 7$ **3** $f(x) = 3x^2$ **4** $f(x) = 5x^2$

5 $f(x) = x^3$ **6** $f(x) = 4x^3 + 3x$ **7** $f(x) = \dfrac{2}{x}$ **8** $f(x) = \dfrac{4}{x^2}$

EXERCISE 1B

Find the derivative of each of the following from first principles.

1 $f(x) = \sin 2x$ **2** $f(x) = \sin 3x$ **3** $f(x) = (x - 5)^2$ **4** $f(x) = (2x + 3)^2$

5 $f(x) = \dfrac{1}{2x + 1}$ **6** $f(x) = \dfrac{1}{(x - 3)^2}$ **7** $f(x) = \cos x$ **8** $f(x) = \cos(2x - 1)$

Reminder

Standard derivatives

In Exercises 1A and B you found the derivative from first principles. In practice, we simply want to write down derivatives, if possible. The derivatives in Exercises 1A and 1B could have been written down using rules with which you are already familiar.

If $f(x) = ax^n$ then $f'(x) = nax^{n-1}$

If $f(x) = \sin ax$ then $f'(x) = a\cos ax$

If $f(x) = \cos ax$ then $f'(x) = -a\sin ax$

Another way of writing the rules:

$$\frac{\mathrm{d}}{\mathrm{d}x}(ax^n) = nax^{n-1}$$

$$\frac{\mathrm{d}}{\mathrm{d}x}(\sin ax) = a\cos ax$$

$$\frac{\mathrm{d}}{\mathrm{d}x}(\cos ax) = -a\sin ax$$

While the rules for $\sin x$ and $\cos x$ have been justified in Example 3 and Exercise 1B question 7, the justification for x^n is somewhat more complicated and we will deal with this later on!

You have also used the sum and differences rule, the chain rule and the product rule, but they are covered in the following sections, with more formal justification.

Sums and differences rule

Given that f and g are differentiable and $k(x) = f(x) + g(x)$ then $k'(x) = f'(x) + g'(x)$.

Proof

$$\frac{k(x + h) - k(x)}{h} = \frac{f(x + h) + g(x + h) - [f(x) + g(x)]}{h}$$

$$= \frac{f(x + h) - f(x)}{h} + \frac{g(x + h) - g(x)}{h}$$

$$k'(x) = \lim_{h \to 0}\left[\frac{f(x + h) - f(x)}{h} + \frac{g(x + h) - g(x)}{h}\right]$$

$$= \lim_{h \to 0}\left[\frac{f(x + h) - f(x)}{h}\right] + \lim_{h \to 0}\left[\frac{g(x + h) - g(x)}{h}\right]$$

$$= f'(x) + g'(x)$$

Expressing this result another way:

$$\frac{d}{dx}(f(x) + g(x)) = \frac{d}{dx}f(x) + \frac{d}{dx}g(x)$$

or $\quad y = f(x) + g(x)$

$$\frac{dy}{dx} = \frac{df}{dx} + \frac{dg}{dx}$$

Example

Three ways of asking essentially the same question

If $f(x) = x^2 + \sin x$, find $f'(x)$. 　　　Differentiate $x^2 + \sin x$. 　　　Find the derivative, $\dfrac{dy}{dx}$, of $y = x^2 + \sin x$.

Three ways of writing the answer!

$f'(x) = 2x + \cos x$ 　　　$\dfrac{d}{dx}(x^2 + \sin x) = 2x + \cos x$ 　　　$\dfrac{dy}{dx} = 2x + \cos x$

EXERCISE 2

1 Find the derivatives of: 　**a** $x^3 + \cos x$ 　　　　　　　　**b** $\sin 3x + x^5$

2 Differentiate:

　a $x^{-2} + x^2$ 　　　　**b** $\dfrac{2}{x^4} + 5x^2$ 　　　　**c** $\dfrac{3}{x^3} + \cos 2x$ 　　　　**d** $\sin 4x + \cos 4x$

3 **a** If $f(x) = \dfrac{x^3 - 3x}{x^2}$ find $f'(x)$.

　b Given that $f(x) = \dfrac{2x^2 + 3x}{x^5}$, calculate $f'(x)$.

4 Find $\dfrac{dy}{dx}$ given that:

a $y = \dfrac{3x^3 - 4x^2}{x^2}$

b $y = \dfrac{4(x + 2)^3 + 3(x + 2)^2}{(x + 2)^2}$

Note

Often, more careful manipulation of algebra is required than knowledge of differentiation!

5 Find $f'(1)$ when:

a $f(x) = \dfrac{1}{2x^4} + 5x^2$

b $f(x) = \cos\dfrac{1}{2}x - \dfrac{3}{2x^3}$

Note that, in questions 2 to 5, the functions are not differentiable for all values of x.

Chain rule

Given that g is differentiable at x, f is differentiable at $g(x)$ and $k(x) = f(g(x))$, then k is differentiable and $k'(x) = f'(g(x)).g'(x)$.

Proof
Let $u = g(x)$ and $t = g(x + h) - g(x)$ noting that $t \to 0$ as $h \to 0$.
Hence $g(x + h) = g(x) + t = u + t$.

$$\frac{f(g(x + h)) - f(g(x))}{h} = \frac{f(u + t) - f(u)}{h}$$

$$= \frac{f(u + t) - f(u)}{t} \times \frac{t}{h}$$

$$\lim_{h \to 0}\left[\frac{f(g(x + h)) - f(g(x))}{h}\right] = \lim_{h \to 0}\left[\frac{f(u + t) - f(u)}{h} \times \frac{t}{h}\right]$$

$$= \lim_{h \to 0}\left[\frac{f(u + t) - f(u)}{t}\right] \times \lim_{h \to 0}\left[\frac{t}{h}\right]$$

$$= \lim_{t \to 0}\left[\frac{f(u + t) - f(u)}{t}\right] \times \lim_{h \to 0}\left[\frac{g(x + h) - g(x)}{h}\right]$$

$$= f'(u) \times g'(x)$$

$$k'(x) = f'(g(x)).g'(x)$$

Note

The proof given here assumes that $g(x + h) \neq g(x)$ for some h close to zero. For a complete proof, more advanced works need to be studied.

Written in Leibniz notation:
$$k = f(u) \text{ and } u = g(x)$$
$$\frac{dk}{du} = f'(u), \quad \frac{du}{dx} = g'(x)$$
so $\dfrac{dk}{dx} = \dfrac{dk}{du} \times \dfrac{du}{dx}$

This form is generally quoted as
$$\frac{dy}{dx} = \frac{dy}{du} \times \frac{du}{dx}$$

Example 1   Find $\dfrac{d}{dx}(2x-3)^5$.

Expressing in the form $f(g(x))$ gives:

$$f(x) = (x)^5 \quad \text{and} \quad g(x) = 2x - 3$$
$$f'(x) = 5(x)^4 \quad \text{and} \quad g'(x) = 2$$

so $\dfrac{d}{dx}(f(g(x))) = f'(g(x)).g'(x)$

$$\dfrac{d}{dx}(2x-3)^5 = 5(2x-3)^4 \times 2$$
$$= 10(2x-3)^4$$

> Alternatively, using Leibniz notation:
> $$y = (2x-3)^5$$
> Let $y = u^5$, where $u = 2x - 3$.
> $$\dfrac{dy}{dx} = \dfrac{dy}{du} \times \dfrac{du}{dx}$$
> $$= 5u^4 \times 2$$
> $$= 5(2x-3)^4 \times 2$$
> $$= 10(2x-3)^4$$

After some practice you should be able to identify the different functions mentally and then write down the derivative:

$$\dfrac{d}{dx}(2x-3)^5 = 5(2x-3)^4 \times \dfrac{d}{dx}(2x-3)$$
$$= 5(2x-3)^4 \times 2$$
$$= 10(2x-3)^4$$

Example 2   Find $\dfrac{d}{dx}(\sin(x^2 + 3x))$.

$$\dfrac{d}{dx}(\sin(x^2+3x)) = \cos(x^2+3x).\dfrac{d}{dx}(x^2+3x)$$
$$= \cos(x^2+3x) \times (2x+3)$$
$$= (2x+3)\cos(x^2+3x)$$

> Alternatively, using Leibniz notation:
> $$y = \sin(x^2 + 3x)$$
> Let $y = \sin u$, where $u = x^2 + 3x$.
> $$\dfrac{dy}{dx} = \dfrac{dy}{du} \times \dfrac{du}{dx}$$
> $$= \cos u \times (2x+3)$$
> $$= (2x+3)\cos(x^2+3x)$$

EXERCISE 3A

1 Find the derivative of each of the following.

   a $(5x+1)^3$ **b** $(4x-7)^6$ **c** $(4x^2-7x)^3$ **d** $\sin(x^2)$

2 Find $f'(x)$ given:

   a $f(x) = \cos 2x$ **b** $f(x) = (5 + 2x - 7x^2)^3$ **c** $f(x) = \cos(x^2+3x)$

   d $f(x) = \sin^3 x$ [i.e. $(\sin x)^3$]

3 Find $\dfrac{dy}{dx}$ for each of the following.

   a $y = \dfrac{3}{(1-4x)^2}$ **b** $y = \dfrac{3}{1-4x^2}$ **c** $y = (x^2 + 3x + 1)^4$

4 Differentiate:

   a $\sin x°$ [Hint: convert $x°$ to radians first.]

   b $\cos x°$

   c $\sin(3x - 30)°$

5 Find the derived function when: **a** $f(x) = \dfrac{1}{\sin x}$ **b** $f(x) = \dfrac{1}{\cos x}$

6 Differentiate the following.

   a $\sin(\cos x)$ **b** $\cos(\sin x)$ **c** $\cos(\cos x)$ **d** $\sin(\sin x)$

7 **a** By expressing $y = x$ in the form $y = \sin(\sin^{-1}(x))$ where $-\dfrac{\pi}{2} \le x \le \dfrac{\pi}{2}$, prove that

      $\dfrac{d}{dx}(\sin^{-1}(x)) = \dfrac{1}{\cos(\sin^{-1}(x))}$.

   b Use the identity $\sin^2 x + \cos^2 x = 1$ to express $\cos(\sin^{-1}(x))$ in terms of $\sin(\sin^{-1}(x))$ and simplify.

   c Hence express the derivative of $\sin^{-1}(x)$ in terms of x without the use of trigonometric functions.

   d Differentiate $\cos^{-1}(x)$.

The chain rule can easily be adapted to deal with more complicated functions. For example, if $y = f(g(h(x)))$, then, using Leibniz notation, we have

$$y = f(u) \text{ where } u = g(t) \text{ and } t = h(x) \text{ so } \frac{dy}{dx} = \frac{dy}{du} \times \frac{du}{dt} \times \frac{dt}{dx}$$

Example   Differentiate $y = \cos^2 3x$.

$$y = \cos^2 3x$$
$$y = (u)^2 \text{ where } u = \cos t \text{ and } t = 3x$$
$$\frac{dy}{dx} = \frac{dy}{du} \times \frac{du}{dt} \times \frac{dt}{dx}$$
$$= 2u \times (-\sin t) \times 3$$
$$= -6 \sin t \cos t$$
$$= -6 \sin 3x \cos 3x$$
$$= -3 \sin 6x \qquad\qquad \text{using } \sin 2A = 2 \sin A \cos A$$

EXERCISE 3B

1 Find $f'(x)$ for each of the following.

   a $f(x) = \sin^2 3x$ **b** $f(x) = \cos^2(\sin x)$ **c** $f(x) = (x + \sin 3x)^2$

   d $f(x) = \cos(\sin^2 x)$

2 Differentiate:

   a $\cos^3(2x + 4)$ **b** $\dfrac{1}{\sin^2(3x + 1)}$ **c** $\cos\left(\dfrac{1}{x^2 + 2x + 1}\right)$

3 Find the derivative of:

a $\dfrac{1}{\cos(x^2 + x)}$

b $\dfrac{1}{\sin(\cos x)}$

c $\dfrac{1}{\sqrt{\sin(3x + 2)}}$

4 a Express $\dfrac{\sin^3 x + \cos^2 x}{\sin^2 x}$ as sums or differences of powers of $\sin x$.

[You will need the identity $\sin^2 x + \cos^2 x = 1$.]

b Hence or otherwise differentiate $\dfrac{\sin^3 x + \cos^2 x}{\sin^2 x}$.

Product rule

Given that f and g are differentiable and $k(x) = f(x).g(x)$ then
$k'(x) = f'(x).g(x) + f(x).g'(x)$

Proof

$$\frac{k(x + h) - k(x)}{h} = \frac{f(x + h).g(x + h) - [f(x).g(x)]}{h}$$

$$= \frac{f(x + h).g(x + h) - f(x).g(x + h) + f(x).g(x + h) - f(x).g(x)}{h} \quad \text{adding extra terms for convenience}$$

$$= \frac{[f(x + h) - f(x)].g(x + h) + f(x).[g(x + h) - g(x)]}{h} \quad \text{taking common factor}$$

$$= \frac{f(x + h) - f(x)}{h} g(x + h) + f(x)\frac{g(x + h) - g(x)}{h} \quad \text{separating fractions}$$

$$k'(x) = \lim_{h \to 0}\left[\frac{f(x + h) - f(x)}{h} g(x + h) + f(x)\frac{g(x + h) - g(x)}{h}\right]$$

$$= \lim_{h \to 0}\left[\frac{f(x + h) - f(x)}{h} g(x + h)\right] + \lim_{h \to 0}\left[f(x)\frac{g(x + h) - g(x)}{h}\right]$$

$$= f'(x).g(x) + f(x).g'(x)$$

Using Leibniz notation you can write this result in an abbreviated form:

If $y = f(x)g(x)$ then $\dfrac{dy}{dx} = \dfrac{df}{dx}.g + f.\dfrac{dg}{dx}$

or $y' = f'.g + f.g'$

Example 1   Differentiate $x^2 \sin x$.

Here $f(x) = x^2$, $g(x) = \sin x$, so
$\quad\quad f'(x) = 2x$, $g'(x) = \cos x$

$$\frac{d}{dx}(x^2 \sin x) = 2x \sin x + x^2 \cos x$$

This line is perfectly acceptable as a final answer but in a more involved question it may be useful to write the answer in a factorised form. Hence consider it good

practice to write $\dfrac{d}{dx}(x^2 \sin x) = 2x \sin x + x^2 \cos x$

$$= x(2 \sin x + x \cos x)$$

Example 2   Find the derivative of $(x + 3)^4(x − 3)^5$.

Let $f(x) = (x + 3)^4$ and $g(x) = (x − 3)^5$. Then

$\quad\quad f'(x) = 4(x + 3)^3$ and $g'(x) = 5(x − 3)^4$ $\hspace{4cm}$ using the chain rule

so $\dfrac{d}{dx}((x + 3)^4(x − 3)^5) = f'(x).g(x) + f(x).g'(x)$ $\hspace{2cm}$ using the product rule

$\hspace{3cm} = 4(x + 3)^3(x − 3)^5 + (x + 3)^4 × 5(x − 3)^4$ $\hspace{2cm}$ *

$\hspace{3cm} = (x + 3)^3(x − 3)^4[4(x − 3) + 5(x + 3)]$

$\hspace{3cm} = (x + 3)^3(x − 3)^4(9x + 3)$

$\hspace{3cm} = 3(x + 3)^4(x − 3)^4$

The * marks the end of the differentiation – the rest is the algebra needed to present the answer in as tidy a form as possible.

Example 3   Differentiate $x^2(x + 1)^3 \sin x$.

Here we have a product of three functions. Not to worry – the product rule for two functions can easily be adapted (using the abbreviated notation):

$\quad\quad y = f(x)g(x)h(x)$

$\quad\quad y' = f'.(g.h) + f.(g.h)'$ $\hspace{3cm}$ using the product rule on f times (gh)

$\quad\quad\quad = f'.g.h + f.(g'.h + g.h')$ $\hspace{2.5cm}$ using the product rule on g times h

$\quad\quad\quad = f'.g.h + f.g'.h + f.g.h'$

so

$\quad\quad \dfrac{d}{dx}(x^2(x + 1)^3 \sin x) = 2x(x + 1)^3 \sin x + x^2.3(x + 1)^2 \sin x + x^2(x + 1)^3 \cos x$

$\hspace{4cm} = x(x + 1)^2[2(x + 1) \sin x + 3x \sin x + x(x + 1) \cos x]$

EXERCISE 4A

1 Find the derivative of each of the following.

a $x^3 \sin x$ $\hspace{3cm}$ **b** $(x + 1)^2 \cos x$ $\hspace{2cm}$ **c** $(x + 1)^4(x − 1)^3$

d $(x + 1)^2(x + 7)^5$ $\hspace{2cm}$ **e** $(x − 1)^4 \cos x$

2 a If $f(x) = x^3(x − 1)^2$, find $f'(1)$.

b If $f(x) = (x − 1)^2 \sin x$, find $f'\left(\dfrac{\pi}{2}\right)$.

3 Differentiate each of the following. You may need both the chain rule and the product rule.

a $(2x + 1)^2 \cos x$ $\hspace{2cm}$ **b** $\sin 2x \cos 3x$ $\hspace{2cm}$ **c** $x^4 \sin 3x$

d $(2x + 1)^2(3x − 1)^4$ $\hspace{1.5cm}$ **e** $(x^2 + x) \sin 2x$ $\hspace{1.5cm}$ **f** $(x^2 − 1)(x^3 − 1)$

g $\sin 2x \sin 3x$ $\hspace{2.5cm}$ **h** $x(x^2 + 3x)^3$ $\hspace{2cm}$ **i** $x^4(x^2 + 3x)$

j $\sin x \cos x$

Although both questions **3i** and **3j** are products, you don't really need the product rule! In **3i** we have

$$\frac{d}{dx}(x^4(x^2 + 3x)) = \frac{d}{dx}(x^6 + 3x^5) \qquad \text{multiplying out gives a sum of two functions}$$

$$= 6x^5 + 15x^4$$

In **3j** we can use a trigonometry identity to simplify

$$\frac{d}{dx}(\sin x \cos x) = \frac{d}{dx}\left(\frac{1}{2}\sin 2x\right) \qquad \text{using the identity } \sin 2A = 2\sin A \cos A$$

$$= 2 \times \frac{1}{2}\cos 2x$$

$$= \cos 2x$$

This is rather quicker than

$$\frac{d}{dx}(\sin x \cos x) = (\sin x)'.\cos x + \sin x.(\cos x)'$$

$$= \cos x.\cos x + \sin x.(-\sin x)$$

$$= \cos^2 x - \sin^2 x$$

$$= \cos 2x \qquad \text{using the identity } \cos^2 A - \sin^2 A = \cos 2A$$

EXERCISE 4B

1 Find the derivative of each of the following.

   a $(1 - 5x)^3(1 + 5x)^3$ **b** $\cos 2x \cos 4x$ **c** $x^2(x - 1)^4 \cos x$

   d $(x^2 + 3x + 1)^4$

2 If $f(x) = x \sin^2 x$, find $f'\left(\dfrac{\pi}{6}\right)$.

3 If $f(x) = x \sin x^2$, find $f'\left(\dfrac{\pi}{3}\right)$.

4 If $f(t) = (8t^2 - 5t)(9t^2 + 4)$, find $f'(t)$.

5 a Show that $\dfrac{d}{dz}(z^3 + 3z)^2(z^2 - 1) = 2z(z^2 + 3)(4z^4 + 9z^2 - 3)$.

   b Explain why $\dfrac{d}{dz}(z^3 - 3z)^2.(z^2 - 1) = 0$ only has three solutions.

6 The diagram shows the graph of $x^2y = 4a^2(2a - y)$.
   This classical graph is known as 'The Witch of Agnesi'.

   a Show that $y = \dfrac{8a^3}{x^2 + 4a^2}$.

   b Hence show that the gradient of the tangent at
      $(2a, a)$ is $-\dfrac{1}{2}$.

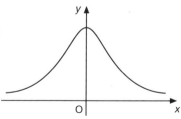

7 If $y = \sin(\cos x).\cos(\cos x)$, show that $\dfrac{dy}{dx} = -\sin x \cos(2 \cos x)$.

Quotient rule

Given that f and g are differentiable and $k(x) = \dfrac{f(x)}{g(x)}$ then

$$k'(x) = \left(\frac{f(x)}{g(x)}\right)' = \frac{f'(x)g(x) - f(x)g'(x)}{(g(x))^2}.$$

Proof

$$k(x) = \frac{f(x)}{g(x)} = f(x) \times (g(x))^{-1} \qquad \text{expressing as a product}$$

$$k'(x) = f'(x) \times (g(x))^{-1} + f(x) \times \frac{\mathrm{d}}{\mathrm{d}x}(g(x))^{-1} \qquad \text{using the product rule}$$

$$= f'(x) \times (g(x))^{-1} + f(x) \times (-1)(g(x))^{-2} \times g'(x) \qquad \text{using the chain rule}$$

$$= \frac{f'(x)}{g(x)} - \frac{f(x)g'(x)}{(g(x))^2}$$

$$= \frac{f'(x)g(x) - f(x)g'(x)}{(g(x))^2}$$

Using Leibniz notation you can write this result in an abbreviated form:

If $y = \dfrac{f(x)}{g(x)}$ then $\dfrac{\mathrm{d}y}{\mathrm{d}x} = \dfrac{\dfrac{\mathrm{d}f}{\mathrm{d}x}\cdot g - f\cdot\dfrac{\mathrm{d}g}{\mathrm{d}x}}{g^2}$

or $\quad y' = \dfrac{f'\cdot g - f\cdot g'}{g^2}$

Example   Find $\dfrac{\mathrm{d}}{\mathrm{d}x}\left(\dfrac{x^3}{\sin x}\right)$.

Here we have $f(x) = x^3$ and $g(x) = \sin x$, so

$$f'(x) = 3x^2 \text{ and } g'(x) = \cos x$$

$$\frac{\mathrm{d}}{\mathrm{d}x}\left(\frac{x^3}{\sin x}\right) = \frac{3x^2 \sin x - x^3 \cos x}{(\sin x)^2}$$

EXERCISE 5A

1 Differentiate the following.

  a $\dfrac{x^4}{x+1}$ **b** $\dfrac{\sin x}{\cos x}$ $(= \tan x)$ **c** $\dfrac{x+1}{x^2+2}$ **d** $\dfrac{\cos x}{x}$

2 Find the derivative of each of the following.

  a $\dfrac{x}{\sin x}$ **b** $\dfrac{3x}{\sqrt{x-3}}$ **c** $\dfrac{\sqrt{x-3}}{3x}$ **d** $\dfrac{x^2}{(x+4)^{\frac{3}{2}}}$

3 Use the quotient rule to differentiate:

  a $\dfrac{1}{\sin x}$ **b** $\dfrac{1}{\cos x}$ **c** $\dfrac{1}{\tan x}$

4 If $f(x) = \dfrac{x+1}{x^2+2}$, find $f'(0)$.   **5** If $f(x) = \dfrac{x^2}{\sqrt{x-1}}$, find $f'(2)$.

37

6 If $y = \dfrac{x^3 - 1}{x^3 + 1}$, show that $\dfrac{dy}{dx} = \dfrac{6x^2 y^2}{(x^3 - 1)^2}$.

7 Solve $\dfrac{dy}{dx} = 0$ where $y = \dfrac{2x^2 + 3x - 6}{x - 2}$.

EXERCISE 5B

1 Differentiate $\dfrac{(3x + 2)^2}{(2x - 1)}$.

2 Find the derivative of $\dfrac{\cos x}{\sin 2x}$.

3 $f(x) = \dfrac{\cos x}{\sin^2 x}$. Find $f'\left(\dfrac{\pi}{4}\right)$.

4 If $y = \dfrac{(x - 1)^3 (x + 2)}{x - 2}$, show that $\dfrac{dy}{dx} = \dfrac{(x - 1)^2 (3x^2 - 4x - 8)}{(x - 2)^2}$.

5 If $f(x) = \dfrac{1}{x^2(x - 1)^3}$, show that $f'(2) = -1$.

A short summary so far

Basic functions:	$\dfrac{d}{dx}(x^n) = nx^{n-1}$ $\dfrac{d}{dx}(\sin x) = \cos x$ $\dfrac{d}{dx}(\cos x) = -\sin x$
Chain rule:	$\dfrac{d}{dx}(f(g(x))) = \dfrac{df}{dg} \times \dfrac{dg}{dx}$
Sums and differences:	$\dfrac{d}{dx}(f \pm g) = f' \pm g'$
Product rule:	$\dfrac{d}{dx}(f.g) = f'.g + f.g'$
Quotient rule:	$\dfrac{d}{dx}\left(\dfrac{f}{g}\right) = \dfrac{f'.g - f.g'}{g^2}$

EXERCISE 6

Miscellaneous

1 Find the derivative of each of the following.

    a $\dfrac{x + 1}{x^2 + 2}$ **b** $\dfrac{1}{\sin 2x}$ **c** $\sin^2 3x$ **d** $(x + 1)^3(x^3 + 1)$

2 If $y = \sin(\sin x)$, find $\dfrac{dy}{dx}$.

3 Find $f'(x)$ when $f(x) = (x + 1)^2 \cos 2x$.

4 $y = \dfrac{\cos x}{\cos x + \sin x}$. Show that $\dfrac{dy}{dx} = -\dfrac{1}{1 + \sin 2x}$ and hence find the gradient of the tangent at $x = \dfrac{\pi}{4}$.

5 Differentiate $\dfrac{(2x + 3)^3}{x^2 + 3x - 1}$.

6 The diagram shows another classical curve, the Serpentine curve, whose equation is $x^2y = ax - a^2y$.

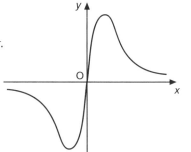

   a Show that $\dfrac{dy}{dx} = \dfrac{a(a^2 - x^2)}{(a^2 + x^2)^2}$.

   b Let m_1, m_2 and m_3 be the gradients of the tangents at $x = \frac{1}{2}a$, $x = a$ and $x = 2a$ respectively. Show that $m_1 + m_2 + 4m_3 = 0$.

Derivatives of sec x, cosec x, cot x, tan x

We start by introducing three 'new' trigonometric functions:

$$f(x) = \sec x \left(\text{where } \sec x = \frac{1}{\cos x}\right) \qquad \text{the secant of } x$$

$$f(x) = \operatorname{cosec} x \left(\text{where } \operatorname{cosec} x = \frac{1}{\sin x}\right) \qquad \text{the cosecant of } x$$

$$f(x) = \cot x \left(\text{where } \cot x = \frac{1}{\tan x}\right) \qquad \text{the cotangent of } x$$

Unlike the sine and cosine functions, the graphs of sec and cosec functions have 'breaks' in them. The functions are otherwise continuous but, at certain values of x, $\sec x$ and $\operatorname{cosec} x$ are undefined. For example, $\sec\left(\dfrac{\pi}{2}\right) = \dfrac{1}{\cos\left(\dfrac{\pi}{2}\right)} = \dfrac{1}{0}$ so $f(x) = \sec x$ is undefined at $x = \dfrac{\pi}{2}$.

In general $\sec x$ is undefined for $x = \dfrac{\pi}{2} + n\pi$ and $\operatorname{cosec} x$ for $x = n\pi$.

   a Have a look at the graphs of these functions on your calculator or set up a table of values in a spreadsheet

	A	B
1	0	=1/cos(A1)
2	=A1+PI()/30	=1/cos(A2)
3	=A2+PI()/30	=1/cos(A3)
4	...	...

	A	B
1	0	1
2	0.105	1.006
3	...	...

This table shows the fomulae required for the graph of $y = \sec x$. Note that PI() = π.

This table shows the start values obtained to give the graph of $y = \sec x$ for $0 \le x \le 2\pi$.

You should obtain the following pictures:

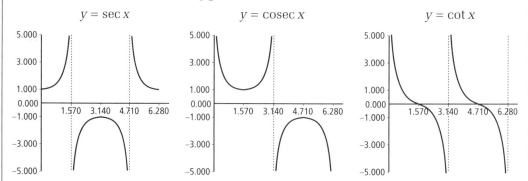

$$y = \sec x \qquad\qquad y = \operatorname{cosec} x \qquad\qquad y = \cot x$$

Note that both the spreadsheet and the graphics calculator try to join up the separate parts of the curve.
You have to learn to ignore these apparent joins.

Note also that the spreadsheet is unable to print $\dfrac{\pi}{2}$, π etc. along the x-axis although it will accept π (entered as PI()) as part of a formula as shown in the formula table above.

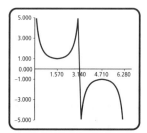

Example   Find the derivative of $\tan x$.

$$\frac{\mathrm{d}}{\mathrm{d}x}\tan x = \frac{\mathrm{d}}{\mathrm{d}x}\left(\frac{\sin x}{\cos x}\right)$$

$$= \frac{\cos x.\cos x - \sin x.(-\sin x)}{\cos^2 x} \qquad \text{using the quotient rule}$$

$$= \frac{\cos^2 x + \sin^2 x}{\cos^2 x}$$

$$= \frac{1}{\cos^2 x}$$

$$= \sec^2 x \qquad \text{no fractional form}$$

EXERCISE 7

1 **a** Use the chain rule to show that the derivative of $\sec x$ is $\sec x \tan x$.
   b Use the quotient rule to show that the derivative of $\operatorname{cosec} x = -\operatorname{cosec} x \cot x$.
   c Show that the derivative of $\cot x$ is $-\operatorname{cosec}^2 x$.

2 Find the derivative of each of the following.

a $\sec 2x$	**b** $\tan 3x$	**c** $\operatorname{cosec} ax$	**d** $\operatorname{cosec}(2x + 3)$
e $\sec(4 - 3x^2)$	**f** $\cot 5x$	**g** $\cot(x^2)$	**h** $\tan(1 - 17x)$

3 Calculate $\dfrac{\mathrm{d}y}{\mathrm{d}x}$ in each case.

a $y = \sec x \tan x$	**b** $y = \cot(\tan x)$	**c** $y = \operatorname{cosec}(\sin x)$
d $y = \operatorname{cosec}^2 3x$	**e** $y = \sec^2 x$	**f** $y = \tan^2 4x$

g $y = \sqrt{\sec x}$

h $y = \dfrac{1}{\sqrt{(1 + \operatorname{cosec} x)}}$

4 Find $f'(x)$ when:

a $f(x) = \dfrac{x^2 + x}{1 + \cot x}$

b $f(x) = \dfrac{\cot x + \sec x}{\cot x - \sec x}$

c $f(x) = \dfrac{\sec x + \cot x}{x^2 + 2x + 1}$

5 Given that $f(x) = \sin^2 x \tan x$, show that $f'\left(\dfrac{\pi}{4}\right) = 2$.

6 If $f(x) = \sin x \sec x$, show that $f'\left(\dfrac{\pi}{3}\right) = 4$.

> **Note**
>
> The function in question **6** would have been more easily differentiated if it had been simplified first. The correct answer, however, is obtained by either method.

Exponential and logarithmic functions

The natural exponential function, e^x, and its inverse $\ln x$, the natural logarithmic function, are perhaps the most important functions in applications of calculus in the real world.

Forms of the exponential function occur, for instance, in:

- probability $f(x) = e^{-\frac{1}{2}x^2}$,
- radioactive decay $q(t) = q_0 e^{-ct}$,
- traffic control studies $f(x) = cx^n e^{-ax}$

Note that $f(x) = e^x$ can be written as $f(x) = \exp x$.

Differentiating exponential functions

LAMD

Let $f(x) = a^x$

$$f'(x) = \lim_{h \to 0}\left[\frac{f(x + h) - f(x)}{h}\right]$$

$$= \lim_{h \to 0}\left[\frac{a^{x+h} - a^x}{h}\right]$$

$$= \lim_{h \to 0}\left[\frac{a^x(a^h - 1)}{h}\right]$$

$$= \lim_{h \to 0}[a^x] \times \lim_{h \to 0}\left[\frac{a^h - 1}{h}\right]$$

$$= a^x \times \lim_{h \to 0}\left[\frac{a^h - 1}{h}\right]$$

With the aid of a calculator or a spreadsheet explore the limit

$$\lim_{h \to 0}\left[\frac{a^h - 1}{h}\right]$$ for different values of a.

On a spreadsheet the table of equations may start with:

	A	B	C	D
1			\multicolumn{2}{c}{h}	
2			0.1	0.01
3	a	2	=($B3^C$2-1)/C$2	=($B3^D$2-1)/D$2

The corresponding table of values would then look like this:

	A	B	C	D	E	F
1			\multicolumn{4}{c}{h}			
2			0.1	0.01	0.001	0.0001
3	a	2	0.72	0.696	0.6934	0.69317
4		3	1.16	1.105	1.0992	1.09867

- Try different values of a.
- Which value of a produces the simplest limit, 1?

The value of a for which $\lim_{h \to 0}\left[\dfrac{a^h - 1}{h}\right] = 1$ is denoted by the symbol e.

It was studied by the mathematician Leonard Euler who first used this symbol. [The value of e, correct to nine decimal places, is 2.718 281 828.]

> Thus if $f(x) = e^x$ then $f'(x) = e^x$.

Example 1   Find the derivative of e^{3x}.

$$\frac{\mathrm{d}e^{3x}}{\mathrm{d}x} = e^{3x} \times 3 \qquad\qquad \text{using the chain rule}$$

$$= 3e^{3x}$$

Example 2   Find the derivative of xe^x.

$$\frac{\mathrm{d}xe^x}{\mathrm{d}x} = 1.e^x + xe^x \qquad\qquad \text{using the product rule}$$

$$= e^x(1 + x)$$

Differentiating the logarithmic function

The natural logarithmic function is written as $f(x) = \ln x$ (or $\log_e x$).

Examples of situations which involve this function are:

- height h is a function of age $h(x) = a + bx + c \ln x$
- age T of an animal is a function of its length L $T(x) = -a \ln ((b - L)/c)$
- weight W of a child is a function of his/her height $W(h) = \ln a + bh$

The natural logarithmic function is the inverse of the exponential function so $e^{\ln x} = x$.

Differentiating with respect to x

$$\frac{de^{\ln x}}{dx} = \frac{dx}{dx}$$

$$e^{\ln x} \frac{d \ln x}{dx} = 1 \qquad \qquad \text{using the chain rule}$$

$$\frac{d \ln x}{dx} = \frac{1}{e^{\ln x}}$$

$$\boxed{\frac{d \ln x}{dx} = \frac{1}{x}}$$

Example 1   Find the derivative of $\ln 3x$.

$$\frac{d \ln 3x}{dx} = \frac{1}{3x} \times 3 \qquad \qquad \text{using the chain rule}$$

$$= \frac{1}{x}$$

Example 2   Differentiate $\dfrac{\ln x}{x^2}$.

$$\frac{d}{dx}\left(\frac{\ln x}{x^2}\right) = \frac{\frac{1}{x}x^2 - \ln x . 2x}{(x^2)^2} \qquad \qquad \text{using the quotient rule}$$

$$= \frac{x(1 - 2\ln x)}{x^4}$$

$$= \frac{1 - \ln x^2}{x^3}$$

EXERCISE 8A

1 Find the derivative of each of the following.

   a e^{5x} **b** $\ln (x + 5)$ **c** e^{7x^2} **d** $\ln x^6$ **e** $e^{\sin x}$

2 Differentiate:

   a $e^{-\frac{1}{x}}$ **b** $\ln (4x^3 - 1)$ **c** $\ln (\tan x)$ **d** $\tan (\ln x)$ **e** $\ln (e^x)$

3 Calculate $f'(x)$ when $f(x)$ is:

   a $\ln(\sin 2x)$ **b** $e^{\ln(3x)}$ **c** $\ln(\ln x)$ **d** $e^{\tan 3x}$ **e** e^{e^x}

4 Find $\dfrac{dy}{dx}$ when y is:

   a $(2x+4)e^{(x+2)}$ **b** $3x\,e^x$ **c** $e^{2x}\sin x$ **d** $e^x\ln x$

   e $x\ln x$

5 Differentiate, using the quotient rule:

   a $\dfrac{x}{\ln x}$ **b** $\dfrac{e^{2x}}{2x}$ **c** $\dfrac{3\ln x}{3x+2}$ **d** $\dfrac{x^2+3x+1}{e^x}$

   e $\dfrac{\ln(3x-1)}{e^{(2x-4)}}$

6 Differentiate:

   a $e^{2x}\cos(2x+1)$ **b** $(\ln x)^2\,e^x$ **c** $\sin\!\left(\dfrac{\ln x}{2x}\right)$ **d** $\dfrac{e^{\sin x}}{\sin x}$

   e $\dfrac{\ln(\cos x)}{x^2}$

In questions **2e** and **3b** above, you could have made use of the fact that $f(f^{-1}(x)) = x$. Remember that $y = a^x \Leftrightarrow \log_a y = x$.

EXERCISE 8B

1 **a** State why you know that $e^{\ln 3} = 3$.

   b Show that $3^x = e^{x\ln 3}$.

   c Hence differentiate 3^x.

   d In a similar fashion find the derivative of:

      (i) 4^x **(ii)** 6^x **(iii)** a^x **(iv)** $4^{(2x+1)}$

   e Show that if $f(x) = x^x$ then $f'(x) = (1 + \ln x)x^x$.

2 $y = \log_{10} x$

   a Make x the subject of the formula.

   b Take the natural log of both sides and simplify using the laws of logs.

   c Make y the subject of the ensuing formula.

   d Hence find the derivative of $\log_{10} x$.

3 It is a known fact that, for any positive integer, a, $\log_a x = \dfrac{\ln x}{\ln a}$.

   Use this fact to differentiate:

   a $\log_2 x$ **b** $\log_5 2x$ **c** $\log_{10}(3x+5)$

Higher derivatives

Given that a function f is differentiable, if f' is also differentiable, then its derivative is denoted by f''; it is called the second derivative of f.

The two notations are:

function	1st derivative	2nd derivative	...	nth derivative
f	f'	f''	$\cdots$	$f^{(n)}$
	$\dfrac{\mathrm{d}f}{\mathrm{d}x}$	$\dfrac{\mathrm{d}^2 f}{\mathrm{d}x^2}$	$\cdots$	$\dfrac{\mathrm{d}^n f}{\mathrm{d}x^n}$

Example

If $y = 3x^4$ then

$$\frac{\mathrm{d}y}{\mathrm{d}x} = 12x^3$$

$$\frac{\mathrm{d}^2 y}{\mathrm{d}x^2} = 36x^2$$

$$\frac{\mathrm{d}^3 y}{\mathrm{d}x^3} = 72x$$

$$\frac{\mathrm{d}^4 y}{\mathrm{d}x^4} = 72$$

$$\frac{\mathrm{d}^5 y}{\mathrm{d}x^5} = 0$$

If $f(x) = \sin x$ then

$$f'(x) = \cos x$$

$$f''(x) = -\sin x$$

$$f'''(x) = -\cos x$$

$$f''''(x) = \sin x$$

$$etc.$$

Continual differentiation of any polynomial, as in the first case of the above example, will eventually lead to 0. In the example, the nth derivative is zero for $n > 4$. However for the trigonometric function a pattern appears to exist for successive derivatives.

$$f(x) = \sin x$$

$$f'(x) = \cos x = \sin\left(x + \frac{\pi}{2}\right)$$

$$f''(x) = \cos\left(x + \frac{\pi}{2}\right) = \sin\left(x + \frac{\pi}{2} + \frac{\pi}{2}\right) = \sin(x + \pi)$$

$$f'''(x) = \cos(x + \pi) = \sin\left(x + \pi + \frac{\pi}{2}\right) = \sin\left(x + \frac{3\pi}{2}\right)$$

At this stage we may make a conjecture that

$$\text{if } f(x) = \sin x$$

$$\text{then } f^{(n)}(x) = \sin\left(x + \frac{n\pi}{2}\right)$$

The proof of this is another matter altogether, requiring the technique of induction which you will meet later on.

EXERCISE 9A

1 a For each of the following functions go through the process of continual differentiation until a constant is achieved.

(i) x^2　　(ii) x^3　　(iii) x^4　　(iv) x^5

b Write down a formula for the value of the constant when $y = x^n$.

2 a Write down the first and second derivatives of $(2x + 1)^3$.

b What is the lowest value of n for which $f^{(n)}(x) = 0$?

3 Which derivatives do not equal zero for $(4x + 3)^4$?

4 For each of the following functions, **(i)** write down its first, second and third derivative, **(ii)** make a conjecture about its nth derivative.

a $\cos 2x$ 　　　　　　　　**b** x^{-1} [Hint: $(-1)^n$ is 1 when n is even and -1 when n is odd.]

c $\ln x$ 　　　　　　　　　**d** e^{4x} 　　　　　　　　**e** $\sqrt{x}$ 　　　　　　　　**f** $x\,e^x$

5 a If $y = \tan x$, find $\dfrac{dy}{dx}$ and $\dfrac{d^2y}{dx^2}$.

b If $y = \ln(\cos x)$, find $\dfrac{dy}{dx}$ and $\dfrac{d^2y}{dx^2}$.

c Write down a connection between the derivatives of $y = \ln(\cos x)$ and $y = \tan x$.

6 If $y = e^{ax}$ where a is a constant, make a conjecture about $\dfrac{d^n y}{dx^n}$.

7 Find the first and second derivatives of $\dfrac{x}{2x + 1}$.

8 If $y = e^x \sin x$ show that $\dfrac{d^4 y}{dx^4} = -4y$.

EXERCISE 9B

1 $y = \sqrt[3]{(x - 1)^4}$.

a Find $\dfrac{dy}{dx}$ and $\dfrac{d^2y}{dx^2}$.

b Use a graphics calculator or spreadsheet to sketch the graphs of y, $\dfrac{dy}{dx}$ and $\dfrac{d^2y}{dx^2}$ for $-2 \le x \le 3$.

c Examine the three graphs and comment on their continuity.

Note that, for the spreadsheet, you will need to enter the equation of the graph as $y = ((x - 1)^4)^{(1/3)}$.

Comment on question 1

Using a spreadsheet you may have a table of equations

	A	B	C	D
1	-2.00	=((A1-1)^4)^(1/3)	=SIGN(A1)*(4/3)*(ABS(A1-1)^(1/3))	=(4/9)((A1-1)^(-2))^(1/3)
2	=A1+0.1	=((A2-1)^4)^(1/3)	=SIGN(A2)*(4/3)*(ABS(A2-1)^(1/3))	=(4/9)((A2-1)^(-2))^(1/3)
3	...	...		

and a corresponding table of values

	A	B	C	D
1	-2.00	4.33	-1.92	0.21
2	-1.95	4.14	-1.90	0.22
3	...	...		

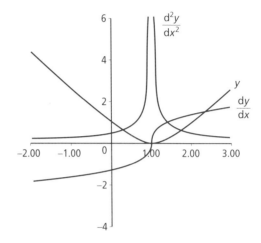

As we can see from the graph, both y and $\dfrac{dy}{dx}$ are continuous. Although it does not show too clearly from the spreadsheet graph, the tangent at $x = 1$ to the graph of $\dfrac{dy}{dx}$ is vertical, i.e. the gradient at $x = 1$ is not defined. This should tell us to expect a discontinuity on the graph of $\dfrac{d^2y}{dx^2}$, and this we can clearly see.

The point being made here is that, just because a function is continuous, this does not mean that all of its derivatives are also continuous.

2 $y = x^{\frac{3}{2}}$

By finding $\dfrac{dy}{dx}, \dfrac{d^2y}{dx^2}$, etc. and sketching the graphs, find the lowest value of n for which $f^{(n)}(x)$ is not continuous.

3 $y = x \ln x - x$

By finding $\dfrac{dy}{dx}, \dfrac{d^2y}{dx^2}$, etc. and sketching the graphs, find the lowest value of n for which $f^{(n)}(x)$ is not continuous.

CHAPTER 2.1 REVIEW

1   Find the derivative of $4x^2 - 3x$ from first principles.

2   Differentiate $\dfrac{4x^5 - 3x}{2x^2}$.

3   If $f(x) = \cos 2x$, find $f'\left(\dfrac{2\pi}{3}\right)$.

4   Differentiate $\sqrt{x^3}.\sin x$.

5   Find the derivative of $\dfrac{\cos x}{(3x-2)^2}$.

6   If $f(x) = \cot 3x$, find $f'(x)$.

7   Find $\dfrac{dy}{dx}$ when $y = \sec x \tan x$.

8   If $y = \operatorname{cosec}^3 x$, show that $\dfrac{dy}{dx} + 3y \cot x = 0$.

9   If $y = x^2 \ln x$, show that $\dfrac{dy}{dx} = x \ln(ex^2)$.

10 $y = e^{t^2}$. Show that $\dfrac{d^2y}{dt^2} - 2t\dfrac{dy}{dt} - 2y = 0$.

CHAPTER 2.1 SUMMARY

1 Definition: the derivative of $f(x)$ is $\lim_{h \to 0}\left(\dfrac{f(x + h) - f(x)}{h}\right)$.

Notation: if $y = f(x)$, then the derivative can be denoted by $\dfrac{dy}{dx}, \dfrac{df}{dx}$ or $f'(x)$ or even simply f'.

nth derivative: if $y = f(x)$, then in particular the second derivative is $\dfrac{d}{dx}\left(\dfrac{dy}{dx}\right) = \dfrac{d^2y}{dx^2}$ and the nth derivative is denoted by $\dfrac{d^ny}{dx^n}$.

2 Basic derivatives

$$\frac{d}{dx}(x^n) = nx^{n-1} \text{ for all real } n, n \neq -1, \quad \frac{d}{dx}(\sin x) = \cos x, \quad \frac{d}{dx}(\cos x) = -\sin x$$

3 Rules

Linear sums: $y = f(x) \pm g(x)$ $\Rightarrow$ $\dfrac{dy}{dx} = \dfrac{df}{dx} \pm \dfrac{dg}{dx}$
$$= f'(x) \pm g'(x)$$

Chain rule: $y = f(g(x))$
i.e. $y = f(u), u = g(x)$ $\Rightarrow$ $\dfrac{dy}{dx} = \dfrac{dy}{du} \cdot \dfrac{du}{dx}$
$$= f'(u).g'(x)$$
$$= f'(g(x)).g'(x)$$

Product rule: $y = f(x).g(x)$ $\Rightarrow$ $\dfrac{dy}{dx} = \dfrac{df}{dx} \cdot g(x) + f(x).\dfrac{dg}{dx}$
$$= f'(x).g(x) + f(x).g'(x)$$

Quotient rule: $y = \dfrac{f(x)}{g(x)}$ $\Rightarrow$ $\dfrac{dy}{dx} = \dfrac{\dfrac{df}{dx} \cdot g(x) - f(x).\dfrac{dg}{dx}}{(g(x))^2}$
$$= \frac{f'(x).g(x) - f(x).g'(x)}{(g(x))^2}$$

4 Further derivatives

$$\frac{d}{dx}(\tan x) = \sec^2 x \qquad\qquad \frac{d}{dx}(\sec x) = \sec x \tan x$$

$$\frac{d}{dx}(\operatorname{cosec} x) = -\operatorname{cosec} x \cot x \qquad\qquad \frac{d}{dx}(\cot x) = -\operatorname{cosec}^2 x$$

$$\frac{d}{dx}(e^x) = e^x \qquad\qquad \frac{d}{dx}(\ln x) = \frac{1}{x}$$

2.2 Applications of Differential Calculus

Historical note

Galileo Galilei

Rectilinear motion, motion in a straight line, was studied by the philosophers of ancient Greece. It was, however, in the early seventeenth century that Galileo did his famous experiments on *free fall*, defining acceleration as

$$a = \frac{V - V_0}{t}$$

where t is the time it takes the velocity to change from V_0 to V. This gave us the formulae $V = V_0 + at$ and $s = V_0 t + \frac{1}{2}at^2$.
It was to explore and develop such concepts that Newton invented the calculus.

Rectilinear motion

A body moves in a straight line along the x-axis.
Its distance, or displacement, from the origin after a time t is x.

If *displacement from the origin* is a function of *time* i.e. $x = f(t)$
then we can derive formulae for velocity, v, and acceleration, a, at time t.

Velocity is the rate of change of *displacement* with *time*: $v = \dfrac{dx}{dt}$

Acceleration is the rate of change of *velocity* with *time*: $a = \dfrac{dv}{dt} = \dfrac{d^2x}{dt^2}$

Note
The use of units must be consistent.

Example 1   A particle travels along the x-axis such that $x(t) = 4t^3 - 2t + 5$ where x represents its displacement in metres from the origin t seconds after observations began.
a How far from the origin is the particle at the start of observations?
b Calculate the velocity and acceleration of the particle after 3 seconds.

a When $t = 0$, $x(0) = 5$ metres.

b $v = \dfrac{dx}{dt} = 12t^2 - 2$.   At $t = 3$, $v = 12 \times 9 - 2 = 106$ m/s

$a = \dfrac{d^2x}{dt^2} = 24t$.   At $t = 3$, $a = 24 \times 3 = 72$ m/s^2

Example 2   A body travels along a straight line such that $s = t^3 - 6t^2 + 9t + 1$ where s represents its displacement in metres from the origin t seconds after observations began.

a Find when **(i)** the velocity is zero   **(ii)** the acceleration is zero.
b When is the distance, s, increasing?
c When is the velocity of the body decreasing?
d Describe the motion of the particle during the first 4 seconds of observations.

a (i) $v = \dfrac{dx}{dt} = 3t^2 - 12t + 9.$ $3t^2 - 12t + 9 = 0 \Rightarrow 3(t - 1)(t - 3) = 0 \Rightarrow t = 1$ or $t = 3.$

Velocity is zero at the first second and the third second.

(ii) $a = \dfrac{d^2x}{dt^2} = 6t - 12.$ $6t - 12 = 0 \Rightarrow t = 2.$ Acceleration is zero at the second second.

b s increasing $\Rightarrow v > 0.$ $3(t - 1)(t - 3) > 0 \Rightarrow t < 1$ or $t > 3.$

c v decreasing $\Rightarrow a < 0.$ $6t - 12 < 0 \Rightarrow t < 2.$

d At $t = 0$ the particle is 1 m from the origin with a velocity of 9 m/s, decelerating at a rate of 12 m/s².

At $t = 1$ it is 5 m from the origin, at rest, decelerating at a rate of 6 m/s².

[At $1 < t < 3$ the velocity is negative; the particle has reversed its direction of motion.]

At $t = 2$ it is 3 m from the origin, its velocity is −3 m/s, with zero acceleration.

At $t = 3$ it is 1 m from the origin, at rest, accelerating at a rate of 3 m/s².

At $t = 4$ it is 5 m from the origin, velocity of 9 m/s, accelerating at a rate of 12 m/s².

Unless otherwise stated, displacement is measured in metres (m), time in seconds (s), velocity in metres per second (m/s or m s^{-1}) and acceleration in metres per second per second (m/s² or m s^{-2}).

EXERCISE 1

1 Given that particles are moving in a straight line according to the equations given below, calculate for each **(i)** the velocity and acceleration after t seconds and **(ii)** the velocity and acceleration after 2 seconds.

a $x = 4t^2 - 6t$

b $x = \dfrac{t}{t + 1}$

c $x = \sqrt[3]{(3t + 2)}$

d $x = 30 \sin 15t$

e $x = t + t^{-1}$

f $x = t^2 + t^{-2}$

2 Find the positions and acceleration of a particle when it comes to rest given that its equation of rectilinear motion is:

a $x = 120t - 16t^2$

b $x = t^3 - 6t^2 + 9t - 1$

c $x = 5t + \dfrac{20}{t + 1}$

d $x = 10 \sin 12t$

e $x = 2t + 8t^{-1}$

f $x = 10t - e^t$

3 The equation of rectilinear motion of a particle is $x = \sqrt{(t + 1)}$.
   a Show that the acceleration is always negative (i.e. the particle is always decelerating).
   b Show that the acceleration is proportional to the cube of the velocity.

4 A projectile is fired straight up with an initial velocity of 300 m/s. Its height, h m, above the ground after t seconds is given by $h(t) = -16t^2 + 300t$.
   a Find (i) the maximum height and   (ii) the time when this is attained.
   b (i) Calculate the velocity with which it hits the ground.
      (ii) Comment on the acceleration at this instant.

5 A car is tested to destruction by accelerating it into a wall on a test site. The test bed is a fixed distance on a straight track. The equation of motion of the car is
   $x = 24t^{\frac{3}{2}}$ where x metres is the distance travelled t seconds into the run.
   The car hits the wall after 4 seconds.
   a How far from the wall did the car start its run?
   b What is its velocity at the moment of impact?
   c What is its acceleration (i) after 1 second   (ii) immediately before impact?

6 A particle oscillates back and forth between two points, A and B. Its distance s metres, measured from a fixed point, at a time $5t$ seconds is given by $s(t) = 8 \sin\left(\dfrac{\pi}{6}t\right)$.
   a Determine the velocity and acceleration of the particle at time $t = 0$.
   b (i) Determine the velocity at point A and at point B.
      (ii) Find the distance between A and B.

7 Viewed from the side, a carriage on a ferris wheel moves with
   rectilinear motion with equation 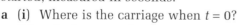 $x = 5 \sin\left(\dfrac{\pi}{30}t\right) + 10$ where x is
   the height above the ground and t is the time since the ride started, measured in seconds.
   a (i) Where is the carriage when $t = 0$?
      (ii) Calculate its velocity and acceleration at this point.
   b How long does it take to make 1 revolution of the wheel?
   c What is the acceleration of the carriage when it is at a height of 12.5 m and rising?

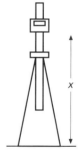

8 As the tide comes in and out, the depth, d metres, of water after t hours is given by
   $d = 3 \sin\left(\dfrac{\pi}{6}t - \dfrac{5\pi}{3}\right) + 4$. At midnight $t = 0$. A buoy floats on the water, rising and falling with the tide.
   a Find the velocity of the buoy at 6 a.m.
   b Determine (i) when the velocity of the buoy is zero
               (ii) the acceleration at these times.

9

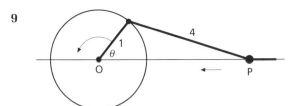

The point P is the end of a crank shaft. As the wheel rotates, P moves back and forward along a straight track. The distance x metres from the origin (the centre of the wheel) after t seconds is given by $x = \cos\theta + \sqrt{(16 - \sin^2\theta)}$ where θ is the angle shown in the diagram. The wheel is rotating at a speed of a quarter of a revolution per second i.e. $\theta = \dfrac{\pi}{2}t$

a Calculate when the velocity of P is zero during the first revolution of the wheel.

b Calculate the velocity of P when $\theta = \dfrac{\pi}{6}$.

c Calculate the acceleration of P when **(i)** $t = 2$ **(ii)** $t = 3$.

10 The height, s metres, reached in t seconds by a body thrown vertically upwards with initial velocity v_1 m s^{-1} is given by the formula $s = v_1 t - \dfrac{1}{2}gt^2$ where g is the acceleration due to gravity.

Find an expression for the maximum height achieved.

11 A meteor entering the atmosphere at time $t = 0$ will approach the Earth with rectilinear motion acording to the equation

$$x = \frac{1}{1 + t^2} + \frac{1}{2}t - 1$$

where x is the distance from the outer atmosphere measured in metres and t is the time in seconds.

a Find an expression for the velocity of the meteor at time t.

b How fast was the meteor travelling after the fifth second?

c As t gets large, the velocity tends to a limit: its terminal velocity. Calculate the terminal velocity.

12 It was suggested that, as a body approached a Black Hole, its apparent motion in a straight line appeared to follow the rule $x = c \ln(t + 1)$ where x units is the distance moved towards the hole since observations began at time $t = 0$ (t measured in seconds) and c is a constant.

a If the body moves 100 units of distance in the 1st second, how far will it move in the 2nd second?

b Discover a formula for **(i)** the velocity   **(ii)** the acceleration after t seconds.

c According to this model, what is the apparent velocity of the falling body after 10 seconds?

Extreme values of a function (extrema)

Critical points

Assuming a is in the domain of a function f, then a *critical point* of the function is any point $(a, f(a))$ where $f'(a) = 0$ or where $f'(a)$ does not exist.

Example　　Consider the piecewise function

$$f(x) = \begin{cases} x^2 & \text{when } -2 \leq x < 1 \\ x & \text{when } 1 \leq x < 2 \\ \frac{1}{2}x + 1 & \text{when } 2 \leq x < 4 \end{cases}$$

defined in the domain [–2, 4) and identify the critical points.

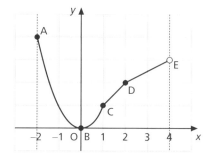

The critical points are:

A(–2, 4)　$f'(-2)$ does not exist; only *right* differentiable at A

B(0, 0)　　$f'(0) = 0$

C(1, 1)　　$f'(1)$ does not exist; *left* derivative = 2, right derivative = 1

D(2, 2)　　$f'(2)$ does not exist; *left* derivative = 1, right derivative = 0.5.

Note that E(4, 3) is not a critical point since 4 is not in the domain of f.

> If $(a, f(a))$ is a *critical point* of the function we will refer to a as a *critical number* and to $f(a)$ as a *critical value* of the function.

Local extreme values (local extrema)

Assuming a is in the domain of a function f, then the function is said to have a *local maximum value, $f(a)$*, at a if, and only if, there exists an interval centred at a in the domain, for which $f(a) \geq f(x)$ for all x in the interval.

Similarly the function is said to have a *local minimum value, $f(a)$*, at a if, and only if, there exists an interval centred at a in the domain, for which $f(a) \leq f(x)$ for all x in the interval.

Example Consider the function

$$f(x) = \begin{cases} x^2 + 2x & \text{when } -3 \le x < -1 \\ x^3 & \text{when } -1 \le x < 1 \\ -2x^2 + 8x - 5 & \text{when } 1 \le x \le 3 \end{cases}$$

Find the critical points and identify any local maximum or minimum values.

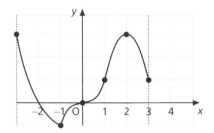

(−1, −1) is a local minimum point,
−1 being the local minimum value.
(2, 3) is a local maximum point,
3 being the local maximum value.

Note that the local extrema all occur at critical points though not all critical points are local extrema. (−3, 3) and (3, 1) are end-points and (0, 0) is a horizontal point of inflexion.

End-point extreme values

If A(a, $f(a)$) is an end-point in the domain then it is said to be

- an *end-point maximum* if there exists a region in the domain for which A is an end-point and for which $f(a) \ge f(x)$ for all x in that region
- an *end-point minimum* if there exists a region in the domain for which A is an end-point and for which $f(a) \le f(x)$ for all x in that region.

Note that end-points are critical points.

Example This function is defined on the domain [−2, 4]. Identify any end-point maximum or minimum.

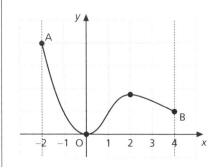

A(−2, 4) is an end-point maximum
B(4, 1) is an end-point minimum.

The nature of stationary points

If $f'(a) = 0$ then a *table of signs* over a suitable interval centred at a provides evidence of the nature of the stationary point that must exist at a.

A simpler test does exist. The derivative at a, $f'(a)$, provides a value for the rate of change of $f(x)$ at $x = a$. In the same way, $f''(a)$ provides a value for the rate of change of $f'(x)$ at $x = a$. Since $f'(x)$ represents the gradient of the function then $f''(x)$ represents the rate of change of the gradient.

The diagram shows, as an example, the graph of the function $f(x) = x^3 - x^2 - 4x + 4$. The numerical values are the values of the gradients at various points on the curve. Around Q (a minimum turning point) the gradient changes from left to right through the values $-3, -2, -1, 0, 1, 2, 3$. At Q the rate of change of the gradient is positive: i.e.

at a minimum turning point $f''(x) > 0$

Around P (a maximum turning point) the gradient changes from left to right through the values $3, 2, 1, 0, -1, -2, -3$. At P the rate of change of the gradient is negative: i.e.

at a maximum turning point $f''(x) < 0$

If the second derivative is easier to determine than making a table of signs then this provides an efficient technique for finding the nature of stationary points.

$f'(x) = 0$ and $f''(x) > 0$	minimum turning point
$f'(x) = 0$ and $f''(x) < 0$	maximum turning point
$f'(x) = 0$ and $f''(x) = 0$	draw a table of signs

EXERCISE 2

1 Examine the graph of each of the piecewise functions and categorise the critical points marked.

a

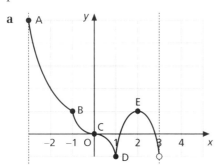

$$f(x) = \begin{cases} x^2 + 2x + 2 & \text{when } -3 \le x < -1 \\ -x^3 & \text{when } -1 \le x < 1 \\ -2x^2 + 8x - 7 & \text{when } 1 \le x < 3 \end{cases}$$

Domain $[-3, 3)$

b

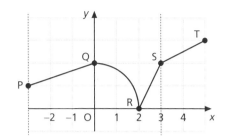

$$f(x) = \begin{cases} \frac{1}{3}x + 2 & \text{when } -3 \leq x < 0 \\ \sqrt{(4 - x^2)} & \text{when } 0 \leq x < 2 \\ 2x - 4 & \text{when } 2 \leq x < 3 \\ \frac{1}{2}(x - 1) & \text{when } 3 \leq x \leq 5 \end{cases}$$

Domain $[-3, 5]$

2 Identify in each of the following functions:
   (i)   where $f'(x) = 0$
   (ii) where $f'(x)$ does not exist, quoting the relevant left and/or right derivatives to reinforce your claim
   (iii) the end-point extrema, stating their natures.

a $f(x) = \begin{cases} 2x^2 - 2x - 2 & \text{when } 0 \leq x < 2 \\ 4 - x & \text{when } 2 \leq x < 4 \\ (x - 4)(x - 8) & \text{when } 4 \leq x \leq 7 \end{cases}$

   Domain $[1, 7]$

b $f(x) = \begin{cases} x + 5 & \text{when } -3 \leq x < -2 \\ x^2 - 1 & \text{when } -2 \leq x < -1 \\ 1 - x^2 & \text{when } -1 \leq x < 1 \\ x^2 - 1 & \text{when } 1 \leq x \leq 2 \end{cases}$

   Domain $[-3, 2]$

c $f(x) = \begin{cases} \sin x & \text{when } 0 \leq x < \dfrac{\pi}{4} \\ \cos x & \text{when } \dfrac{\pi}{4} \leq x < \dfrac{5\pi}{4} \\ \sin x & \text{when } \dfrac{5\pi}{4} \leq x \leq 2\pi \end{cases}$

   Domain $[0, 2\pi]$

d $f(x) = \begin{cases} (x + 2)^2 & \text{when } -3 \leq x < -2 \\ -x^3 & \text{when } -1 \leq x < 1 \\ 3x - 4 & \text{when } 1 \leq x < 2 \end{cases}$

   Domain $[-3, 2)$

3 Investigate the local extrema for the graphs of the following functions.
   Use the second derivative to help you establish their natures.
   Remember: if $f''(x) = 0$, a table of signs is required.
   a $f(x) = x^3 - 9x^2 + 24x - 7$
   b $f(x) = x^3 - 3x^2 - 9x + 11$
   c $f(x) = x^4$
   d $f(x) = x^4 - 8x^2$
   e $f(x) = 2x^5 - 5x^2$
   f $f(x) = (2x - 1)^3$
   g $f(x) = 3x^4 + 8x^3 - 24x^2 - 96x + 10$
   h $f(x) = -x^4 + 8x^3 - 18x^2$
   i $f(x) = x^6 - 3x^4 + 3x^2$

Global extreme values (global extrema)

A point $(a, f(a))$ is said to be a *global maximum* if and only if a is in the domain and $f(a) \geq f(x)$ for all x in the domain. $f(a)$ is referred to as the global maximum value.
A point $(a, f(a))$ is said to be a *global minimum* if and only if a is in the domain and $f(a) \leq f(x)$ for all x in the domain. $f(a)$ is referred to as the global minimum value.

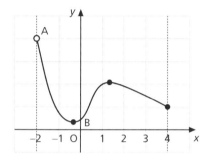

This function is defined on the domain $(-2, 4]$.

It has a global minimum at B.
It has no global maximum.
(A is not in the domain.)

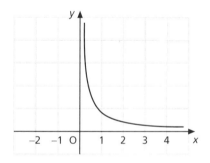

$f(x) = x^{-1}$ in the domain $(0, \infty)$ has no global extrema.
In fact it has no critical points at all.

If a continuous function is defined in a *closed* interval then it must have both a global minimum and a global maximum.

To find them, check the critical points for *local* extrema and *end-point* extrema. The greatest of these must be the *global* maximum and the least of these must be the *global* minimum.

When a function is unbounded then we are guaranteed neither extremes. Care must be taken if the function is not continuous. $\tan x$, for example is undefined at $x = \dfrac{\pi}{2}$ so the domain cannot include $\dfrac{\pi}{2}$. In the domain $\left(-\dfrac{\pi}{2}, \dfrac{\pi}{2}\right)$ $\tan x$ has no global extrema. The one critical point in this domain is a horizontal point of inflexion, so there are no *local* extrema.

The function $f(x) = x^3 - x^2 - 4x + 4$, as discussed on page 56, has a local maximum at P and a local minimum at Q but, since its domain is $(-\infty, \infty)$, it has no global extrema.

Example 1   By considering the critical points of f, find the global extrema of the piecewise function.

$$f(x) = \begin{cases} \dfrac{x^2 + 1}{x} & \text{when } -2 \leq x < 0 \\[2mm] \dfrac{x}{x^2 + 1} & \text{when } 0 \leq x \leq 2 \end{cases} \qquad \text{in the domain } [-2, 2]$$

1 Consider the end-points of each subset of the domain:

In $[0, 2]$ $f(0) = 0$; $f(2) = \frac{2}{5}$;

In $[-2, 0)$ $f(-2) = -\frac{5}{2}$

2 Consider where else $f'(x)$ does not exist, i.e. when $x = 0$.
We must also consider the right-hand side of the $[-2, 0)$ subset:
it is not closed. Note that, as $x \to 0$, $f(x) \to -\infty$.
There will be no global minimum.

3 Consider when $f'(x) = 0$ in each subset of the domain:

Check the derivative is $f'(x) = \begin{cases} \dfrac{x^2 - 1}{x^2} & \text{when } -2 \leq x < 0 \\[2mm] \dfrac{1 - x^2}{(x^2 + 1)^2} & \text{when } 0 \leq x \leq 2 \end{cases}$

In $[0, 2]$ $f'(x) = 0 \Rightarrow 1 - x^2 = 0 \Rightarrow x = 1 \Rightarrow f(1) = \frac{1}{2}$

In $[-2, 0)$ $f'(x) = 0 \Rightarrow x^2 - 1 = 0 \Rightarrow x = -1 \Rightarrow f(-1) = -2$

4 Examine the data:

End-point extrema $f(-2) = -\frac{5}{2}$ and $f(2) = \frac{2}{5}$

Local extrema $f(-1) = -2$ and $f(1) = \frac{1}{2}$

Breaks (discontinuity)   as $x \to 0$ from left, $f \to -\infty$

The global maximum is the greatest of these: $f(2) = \frac{2}{5}$.
There is no global minimum.

Example 2   Sometimes a search cannot be completed by analytical approaches only: find the global maximum and minimum of the function $f(x) = x^2 \sin x$ given that $f(x)$ is a continuous and differentiable function in the interval $0 \leq x \leq 3$.

1 End-points: $f(0) = 0$; $f(3) = 1.27$

2 Continuous and differentiable, so no breaks, joins or cusps: $f'(x)$ exists throughout the domain.

3 Consider when $f'(x) = 0$.
$f'(x) = 2x \sin x + x^2 \cos x = 0 \Rightarrow x(2 \sin x + x \cos x) = 0$
$\Rightarrow x = 0$ or $\tan x + 0.5x = 0$. *Trial and improvement* gives $x = 2.288\,89$ (5 d.p.) and $f(2.288\,89) = 3.945$ (3 d.p.)

4 Examining the data shows us that the global maximum value is $f(2.288\,89) = 3.945$ and the global minimum is $f(0) = 0$.

Newton's method of approximating the roots of an equation

What do we mean when we say that the equation $\tan x + 0.5x = 0$ was solved by *trial and improvement*?

We could use a spreadsheet, entering the formula and searching for a value of x which makes it zero; we could use the table facility offered by many graphics calculators; we could use the graphing and zoom facilities of the graphics calculator.

Another way is to use a dodge invented by Isaac Newton which employs derivatives again. Newton said that if you want to solve the equation $f(x) = 0$ then *guess*! He then suggested that, if x_n is your guess, then x_{n+1} is a better guess where

$$x_{n+1} = x_n - \frac{f(x_n)}{f'(x_n)}$$

When $f(x) = \tan x + 0.5x$ then $f'(x) = \sec^2 x + 0.5$ so $x_{n+1} = x_n - \dfrac{\tan x_n + 0.5x_n}{\sec^2 x_n + 0.5}$

Pick any guess in the domain, say $x_n = 2$.
Put this in the right-hand side of the equation to get $x = 2.188\,869$
Put this in the right-hand side of the equation to get $x = 2.278\,58$
Put this in the right-hand side of the equation to get $x = 2.288\,827$
Put this in the right-hand side of the equation to get $x = 2.288\,929$
The calculator is repeating itself correct to
four decimal places.

This method does not work every time, but it does work often.

> Effort can be minimised with efficient use of the ANS button on your calculator

EXERCISE 3

1 Find the global extrema for the following functions.

    a $f(x) = x^2 - 3x + 2$ in the interval $[1, 3]$ **b** $f(x) = x^2 - 3x + 2$ for $-1 \leq x \leq 2$

    c $f(x) = 4x^3 - x^4$ for $1 \leq x \leq 4$ **d** $f(x) = \dfrac{x}{x^4 + 3}$ for $-2 \leq x \leq 2$

    e $f(x) = \ln(x + 1) - x + 1$ for $0.5 \leq x \leq 1.5$ **f** $f(x) = \dfrac{1}{\sqrt{x^2 + 1}}$ for $-1 \leq x \leq 3$

2 Explore the global extrema of $f(x) = \dfrac{(2x - 3)(3x - 2)}{x}$ when the domain is:

    a $0.5 \leq x \leq 2$ **b** $0 < x \leq 2$

    c $-1 \leq x < 0$ **d** $-2 \leq x \leq -0.5$

3 Explore the global maxima and minima of these functions

   a $f(x) = e^{2x} + 4$ for $-2 \leq x \leq 0$ **b** $f(x) = \ln(3x - 5)$ in the interval $[2, 3]$

4 Find the global extrema of $f(x) = x^2 + x^{-1}$ in the interval:

   a $[-2, 0)$ **b** $(0, 2]$

5 Examine the following piecewise functions for global extrema.

   a $f(x) = \begin{cases} x & \text{when } -1 \leq x \leq 2 \\ 4 - x & \text{when } 2 \leq x \leq 4 \\ x - 1 & \text{when } 4 \leq x \leq 5 \end{cases}$
     b $f(x) = \begin{cases} x(2 - x) & \text{when } -1 \leq x \leq 2 \\ 2 - x & \text{when } 2 \leq x \leq 3 \\ x^2 - 6x - 1 & \text{when } 3 \leq x \leq 5 \end{cases}$

   c $f(x) = \begin{cases} 3^x & \text{when } 0 \leq x \leq 1 \\ 3^{-x+2} & \text{when } 1 \leq x \leq 2 \\ 1 & \text{when } 2 \leq x < 4 \end{cases}$
     d $f(x) = \begin{cases} -x & \text{when } -1 < x \leq 1 \\ x - 2 & \text{when } 1 \leq x \leq 2 \\ \frac{1}{2}x + 1 & \text{when } 2 \leq x < 4 \end{cases}$

                            careful!

   e $f(x) = \begin{cases} -x^{-1} & \text{when } -1 \leq x < 0 \\ x & \text{when } 0 \leq x \leq 2 \\ x - 4 & \text{when } 2 \leq x \leq 3 \end{cases}$
     f $f(x) = \begin{cases} \dfrac{x}{\sqrt{(x)^2}} & \text{when } -5 \leq x < 0 \\ 0 & \text{when } x = 0 \\ \dfrac{x}{\sqrt{(x)^2}} & \text{when } 0 < x \leq 5 \end{cases}$

6 The functions $f(x) = \dfrac{e^x + e^{-x}}{2}$ and $g(x) = \dfrac{e^x - e^{-x}}{2}$ are both important functions in mathematics.

   Examine these functions for critical points and possible global extrema.

   The functions are defined over all real x

7 **a** Find the critical points for the function $f(x) = x \cos x$ in the domain $[0, 2\pi]$.
       Use Newton's method for improving guesses, given the fact that $f'(x) = 0$ close to $x = 1$ and $x = 3$.

   b State the global extrema.

8 Study the function $f(x) = x^2 \ln x$ in the domain $0 < x \leq 2$ for global extrema, given that there is a critical point in the region $0.4 < x < 0.8$.

Optimisation problems

Optimisation problems appear in many guises – often the context in which they are set can be somewhat misleading. Here are some guidelines which you should follow. They should help you make a start and ensure that you do not miss out any key points.

1 Read the question at least twice!

2 Draw a sketch where appropriate. This should help you introduce any variables you are likely to need. It may be that you come back to the diagram to add in an 'x' etc.

3 Try to translate any information in the question into a mathematical statement.

4 Identify which variable is to be optimised (maximised or minimised) and then express this variable as a function of *one* of the other variables.

5 Find the critical numbers for the function arrived at in step 4.

6 Determine the local extrema and (if necessary) the global extrema.

The worked example which follows has the guideline numbers added so that you can identify each step. It is not necessary for you to write these guideline numbers in your solutions.

Example   An open box with a rectangular base is to be constructed from a rectangular piece of card measuring 16 cm by 10 cm. A square is to be cut out from each corner and the sides folded up. Find the size of the cut-out squares so that the resulting box has the largest possible volume.

2 Make a sketch.
   Introduce a variable (x) for the side of each square.

3 The box will have length = $16 - 2x$,
   breadth = $10 - 2x$, height = x.

4 We are concerned with maximising the volume V where $V = x(16 - 2x)(10 - 2x)$ and $0 \leq x \leq 5$. Note that in this example there is only one other variable (x) so we can express V only in terms of x.

5 $V = 4x^3 - 52x^2 + 160x$
   $$\frac{dV}{dx} = 12x^2 - 104x + 160$$
   $$= 4(3x^2 - 26x + 40)$$
   $$= 4(3x - 20)(x - 2)$$
   For critical numbers, $\frac{dV}{dx} = 0$, giving $x = 2$, $x = \frac{20}{3}$ (which is outside the range for x)

6 $\frac{d^2V}{dx^2} = 4(6x - 26)$ *so we have*

x	2
$\dfrac{dV}{dx}$	0
$\dfrac{d^2V}{dx^2}$	−
local:	max

$V(2) = 4 \times 8 - 52 \times 4 + 160 \times 2 = 144$; end-points give $V(0) = 0$ and $V(5) = 0$ so the maximum volume is 144 cm^3.

EXERCISE 4A

1 The turning effect, T, of a power boat, is given by the formula $T = 8 \cos x \sin^2 x$,
$0 < x < \dfrac{\pi}{2}$ where x is the angle (in radians) between the rudder and the central line of
the boat. Find the size of x which maximises the turning effect.

2 An underwater wireless cable consists of a core of copper wires with a casing made
of non-conductive material. Let the ratio of the radius of the core to the thickness of
the casing be t. The speed, V, of the signal is given by $V(t) = kt^2 \ln \dfrac{1}{t}$, $0 < t < 1$,
where k is a constant (>0). Find the value of t which gives maximum speed and the
corresponding maximum speed.

3

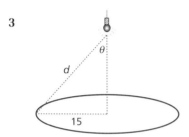

A light is positioned centrally above a circular room of radius 15 feet. The brightness
B at a point on the floor is given by $B = \dfrac{k \cos \theta}{d^2}$ where k is a constant.

At what height should the light be so that the perimeter of the floor is as bright as
possible?

4 A point M lies 500 m north of point N. Cyclist A starts from M, travelling south at
6 m/s. Cyclist B starts from N, moving east at 8 m/s. Let D be the distance between
the two cyclists. Let angle ABN = x.

a Show that $D = \dfrac{4000}{8 \sin x + 6 \cos x}$.

b Find the shortest distance between the cyclists (i.e. the minimum value of D).
[Hint: There is more than one way of tackling **b**. If you use the wave function you
can avoid finding the second derivative, which in this question can be a bit tricky.]

5 Green waste (i.e. grass, leaves, etc.) is taken by private contractors from the Council
depot to a processing plant where it is turned into compost. The lorries cover a
round trip of 60 miles, travelling along roads where the speed limit varies between
40 miles per hour and 60 miles per hour. The plant operators pay the lorry drivers
£12 per hour and reimburse them for their diesel. The diesel costs 80p per litre and
the lorries use up diesel at the rate of $\left(\dfrac{s^2}{300} + 5 \right)$ litres per hour where s is the speed
of the lorry in miles per hour and $s \geq 30$.
At what speed should the lorries travel to minimise the cost to the plant operators?

6 A straight line passes through the point (12, 3) and cuts the axes at A and B. Determine the gradient of the line which minimises OA + OB.

[Hint: since the gradient of the line is related to the angle between the line and the *x*-axis, you should use this angle as your variable i.e. express the distance OA + OB in terms of this angle.]

7 A wall, 2.7 m high, is parallel to, and lies 6.4 m from, the side of a house. A ladder rests on the ground, the top of the wall and against the house. What is the shortest possible length for the ladder?

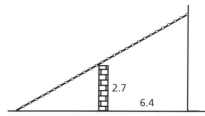

8

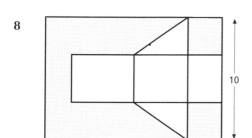

The net shown is to be cut out of a rectangular piece of card, 10 cm wide, and folded up to form a triangular prism.
   a For what value of *x* is the volume a maximum?
   b How long should the piece of card be to keep the waste as small as possible?

9 An open-topped cylinder of volume 12π cm^3 is constructed with the following production costs:

   the base costs 6 pence per square centimetre
   the side costs 6 pence per square centimetre
   the seam round the base costs 3 pence per centimetre.
Determine the dimensions which minimise the cost of production.

EXERCISE 4B

1 The diagram shows a straight line with equation $y = \frac{1}{2}x + 5$ and a parabola with equation $y^2 = 8x$. The point P lies on the parabola and Q lies on the straight line.
   a **(i)** For the distance PQ to be a minimum what can you say about PQ and the line $y = \frac{1}{2}x + 5$?
   (ii) Determine the minimum length of PQ.
   b Show that when the line PQ is in the position of minimum length then PQ is perpendicular to the tangent to the parabola at P.

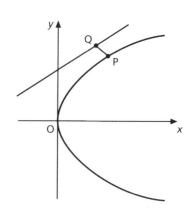

2 A rugby player is positioned at P preparing to kick a penalty between the posts H_1H_2. The player may take up a position anywhere along the line AB. $H_1H_2 = 4$ m and $H_1A = 28$ m. How far back should the player go in order to maximise the angle H_1PH_2?

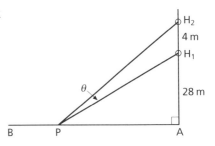

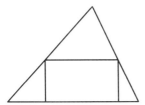

Note
- The maximum distance that the player can kick the ball is 50 metres.
- You will need the tan expansion
$$\tan(A + B) = \frac{\tan A + \tan B}{1 - \tan A \tan B}.$$

3 A farmer has a fixed amount of fencing, F metres, and wishes to fence off a square field of side x metres and a circular field of radius r metres. How should he do this in order to maximise the area fenced off?

4 A cylinder is inscribed in a cone of height h and semi-vertical angle a. Find the maximum volume of the cylinder.

5

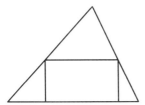

A rectangle is inscribed in a triangle as shown in the diagram.
Determine its maximum area.

6 A cone is formed by cutting a sector out of a circular disc of radius 12 units and then joining the two straight edges. Show that the maximum volume of the cone is 96π cubic units and that the angle of the sector being cut out is $2\pi\left(1 - \sqrt{\frac{2}{3}}\right)$ radians.

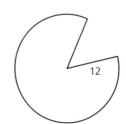

CHAPTER 2.2 REVIEW

1 The equation of rectlinear motion of a particle is given by $x = 8 \ln(t + 1) - 4t$ where x units is the distance from a fixed point at time t seconds.
   Find the acceleration and distance from the fixed point when the velocity is zero.

2 Find the global extrema for the function $f(x) = x(x - 1)^2(x + 1)$.

3 Determine the local extrema for the function $f(x) = x^2(4 - x)^2$.

4 The net of a closed box is to be cut from a piece of card of width L units. The shaded areas in the diagram represent the bits to be cut out (i.e. the wastage).
   a Determine the size of square of side x to be cut out to ensure maximum volume.
   b Find the length of the card in terms of L so that the wastage is kept to a minimum.

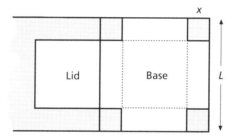

CHAPTER 2.2 SUMMARY

1   If the displacement from the origin, x, of a particle is a function of the time t
    which has elapsed since some arbitrary moment so that $x = f(t)$ then
    (i) the velocity of the particle at time t is given by $v(t) = f'(t)$
    (ii) the acceleration of the particle at time t is given by $a(t) = v'(t) = f''(t)$.
    In Leibniz notation:

$$v = \frac{dx}{dt}$$
$$a = \frac{dv}{dt} = \frac{d^2x}{dt^2}$$

2   If a is in the domain of $f(x)$, then $(a, f(a))$ is a critical point if $f'(a) = 0$ or $f'(a)$
    does not exist.

3   A local maximum occurs at $(a, f(a))$ if there exists a region of the domain centred
    at a where $f(a) \geq f(x)$ for all x in the region.

4   A local minimum occurs at $(a, f(a))$ if there exists a region of the domain centred
    at a where $f(a) \leq f(x)$ for all x in the region.

5   An end-point maximum occurs at $(a, f(a))$ if there exists a region of the domain
    with an end-point at a where $f(a) \geq f(x)$ for all x in the region.

6   A end-point minimum occurs at $(a, f(a))$ if there exists a region of the domain
    with an end-point at a where $f(a) \leq f(x)$ for all x in the region.

7   A global maximum occurs at $(a, f(a))$ if $f(a) \geq f(x)$ for all x in the domain.

8   A global minimum occurs at $(a, f(a))$ if $f(a) \leq f(x)$ for all x in the domain.

9   All extrema are to be found at critical points, though not all critical points need
    be extrema.

10 The derivative can be used to find maxima and minima in practical situations:
    optimisation problems.

3 Integral Calculus

Historical note

Archimedes

While it is true that Isaac Newton and Leibniz are credited with inventing the integral calculus, Archimedes, 287–212 BC, invented the main technique required to prove the fundamental theorem of calculus. His idea was to find the areas of curvilinear shapes by subdividing them into thin strips – so thin that they could be approximated by rectangles.

Using this technique, Archimedes succeeded in finding a formula for the area under a parabola.

Antiderivatives

Reminders

An antiderivative of a function $f(x)$ is a function $F(x)$ where $F'(x) = f(x)$.

For example, $f'(x^3) = 3x^2$ and so x^3 is an antiderivative of $3x^2$.

Of course, any function of the form $x^3 + c$, where c is a constant, is an antiderivative of $3x^2$.

The indefinite integral of antiderivatives of $f(x)$

$$\int f(x)\,dx = F(x) + c$$

represents the family.

Example: $\int 3x^2\,dx = x^3 + c$

The definite integral of $f(x)$ between the limits of a and b can be defined as:

$$\int_a^b f(x)\,dx = F(b) - F(a) \text{ where } F'(x) = f(x)$$

Example:

$$\int_1^2 3x^2\,dx = x^3 + c = \left[x^3\right]_1^2 = 2^3 - 1^3 = 7$$

If $F'(x) = f(x)$ then

$$\int_a^b f(x)\,dx = F(b) - F(a) \text{ is the area under}$$

the curve $y = f(x)$ from $x = a$ to $x = b$.

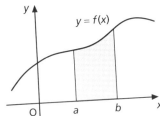

This is often referred to as the *Fundamental Theorem of Calculus.*

$$\int ax^n\,dx = \frac{a}{n+1}\,x^{n+1} + c; \quad n \neq -1$$

$$\int \cos(ax + b)\,dx = \frac{1}{a}\sin(ax + b) + c$$

$$\int \sin(ax + b)\,dx = -\frac{1}{a}\cos(ax + b) + c$$

An outline of the proof of the Fundamental Theorem of Calculus

Consider the following argument for an increasing function, $f(x)$.

Diagram (a) indicates that $A(x)$ represents the area under the curve from the y-axis up to the ordinate through x. The area of the strip shown in Diagram (b) is $A(x + h) - A(x)$ where $h > 0$.

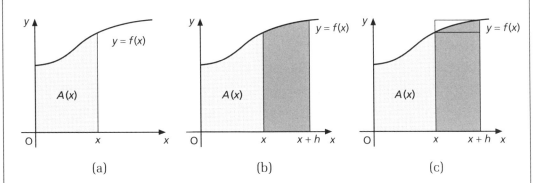

(a)   (b)   (c)

By considering the rectangles in Diagram (c) we see that

$$h \times f(x) \leq A(x + h) - A(x) \leq h \times f(x + h)$$

Dividing throughout by h we get: $\qquad f(x) \leq \left(\dfrac{A(x + h) - A(x)}{h} \right) \leq f(x + h)$

As $h \to 0$, we get: $\qquad f(x) \leq \lim\limits_{h \to 0} \left(\dfrac{A(x + h) - A(x)}{h} \right) \leq f(x)$

$$f(x) \leq A'(x) \leq f(x)$$
$$\Rightarrow \quad f(x) = A'(x)$$
$$A(x) = \int_0^x f(x)\, dx$$

The complete proof would also need to consider continuity and when the function lay under the x-axis. Suffice it to say that this process is only valid for functions which are continuous over the interval in which we integrate.

The following examples have been included to remind you that polynomial functions need to be written as sums and differences before integrating.

Example 1 $\quad \displaystyle\int (x^2 + 3x)^2\, dx = \int (x^4 + 6x^3 + 9x^2)\, dx = \frac{1}{5}x^5 + \frac{3}{2}x^4 + 3x^3 + c$

$$\text{because } \frac{d}{dx}\left(\frac{1}{5}x^5\right) = x^4 \text{ etc.}$$

Example 2 $\quad \displaystyle\int \left(\frac{x - 3}{\sqrt{x}}\right) dx = \int \left(\frac{x}{\sqrt{x}} - \frac{3}{\sqrt{x}}\right) dx = \int \left(x^{\frac{1}{2}} - 3x^{-\frac{1}{2}}\right) dx = \frac{2}{3}x^{\frac{3}{2}} - 6x^{\frac{1}{2}} + c$

$$\text{because } \frac{d}{dx}\left(\frac{2}{3}x^{\frac{3}{2}}\right) = x^{\frac{1}{2}} \text{ etc.}$$

EXERCISE 1A

Calculate the following indefinite integrals.

1 $\int (4x^6)\, dx$ **2** $\int (3x^2 - 7x)\, dx$ **3** $\int (2x^2 - 3)^2\, dx$ **4** $\int (2x - 3)^2\, dx$

5 $\int \left(\dfrac{x^3 - 7}{x^2} \right) dx$ **6** $\int \left(\dfrac{x - 7}{\sqrt{x}} \right) dx$ **7** $\int (\cos 2x)\, dx$ **8** $\int (6 \sin 3x)\, dx$

> **Note**
>
> In question **4** you may have written
>
> $\int (2x - 3)^2\, dx = \int (4x^2 - 12x + 9)\, dx = \frac{4}{3}x^3 - 6x^2 + 9x + c$ whereas the answers
>
> give $\frac{1}{6}(2x - 3)^3 + c$. These are equivalent, only differing by a constant, since
>
> $\frac{1}{6}(2x - 3)^3 + c = \frac{1}{6}(8x^3 - 36x^2 + 54x - 81) + c = \frac{4}{3}x^3 - 6x^2 + 9x - \frac{27}{2} + c$
>
> $\qquad\qquad\qquad\qquad = \frac{4}{3}x^3 - 6x^2 + 9x + c'$

EXERCISE 1B

Find the following indefinite integrals.

1 $\int (8x + 3)^3\, dx$ **2** $\int (5 - 4x)^5\, dx$ **3** $\int (4x - 5)^{-5}\, dx$ **4** $\int (4x^3 - 5)^2\, dx$

5 $\int \sin(3x - 1)\, dx$ **6** $\int \cos(2 - 4x)\, dx$ **7** $\int \cos(1 - x)\, dx$ **8** $\int (3x + 1)^{-2}\, dx$

Four basic rules

1 $\int (af(x) + bg(x))\, dx = a \int f(x)\, dx + b \int g(x)\, dx$

This is simply a statement of a rule that you have been using all along, for example

$\int (x^2 + 2x)\, dx = \int x^2\, dx + 2 \int x\, dx = \frac{1}{3}x^3 + x^2 + c$

2 $\int_a^b f(x)\, dx = F(b) - F(a)$ where $F'(x) = f(x)$, i.e. $F(x)$ is an antiderivative of $f(x)$.

This comes from the definition of a definite integral and the Fundamental Theorem of Calculus.

3 $\int_a^c f(x)\, dx = \int_a^b f(x)\, dx + \int_b^c f(x)\, dx, \ a < b < c$ Example:

A justification: $\int_1^4 3x^2\, dx = \int_1^2 3x^2\, dx + \int_2^4 3x^2\, dx$

$LHS = F(c) - F(a)$ $LHS = \left[x^3 \right]_1^4 = 64 - 1 = 63$

$\qquad = F(b) - F(a) + F(c) - F(b)$ $RHS = \left[x^3 \right]_1^2 + \left[x^3 \right]_2^4$

$\qquad = RHS$ $\qquad = (8 - 1) + (64 - 8) = 63$

4 $\displaystyle\int_a^b f(x)\,dx = -\int_b^a f(x)\,dx$

    A justification:

$$LHS = F(b) - F(a)$$
$$= -(F(a) - F(b))$$
$$= RHS$$

Example:

$$\int_1^4 3x^2\,dx = -\int_4^1 3x^2\,dx$$

$$LHS = \Big[x^3\Big]_1^4 = 64 - 1 = 63$$

$$RHS = \Big[-x^3\Big]_4^1 = -1 - (-64) = 63$$

Standard forms e^x, $\dfrac{1}{x}$, $\sec^2 x$

Since we know $\dfrac{d}{dx}(e^x) = e^x, \ \dfrac{d}{dx}(\ln x) = \dfrac{1}{x}, \ \dfrac{d}{dx}(\tan x) = \sec^2 x$

these give three 'new' antiderivatives.

$$\int (e^x)\,dx = e^x + c, \ \int\left(\frac{1}{x}\right)dx = \ln|x| + c, \ \int (\sec^2 x)\,dx = \tan x + c$$

Note

$|x| = x$ when $x \geq 0$ but $|x| = -x$ when $x \leq 0$.

Example 1   Find $\displaystyle\int_0^1 e^{3x}\,dx + \int_1^2 e^{3x}\,dx$.

$$\int_0^1 e^{3x}\,dx + \int_1^2 e^{3x}\,dx = \int_0^2 e^{3x}\,dx = \left[\frac{1}{3}e^{3x}\right]_0^2 = \frac{1}{3}e^6 - \frac{1}{3} \qquad \text{rules 1, 2 and 3}$$

Example 2   Find $\displaystyle\int \frac{5}{x}\,dx$.

$$\int \frac{5}{x}\,dx = 5\int\left(\frac{1}{x}\right)dx = 5\ln x + c \ (= \ln x^5 + c) \qquad \text{rule 1}$$

Example 3   Find $\displaystyle\int \tan^2 x\,dx$.

First we need some trigonometry.

$$\sin^2 x + \cos^2 x = 1 \Leftrightarrow \frac{\sin^2 x}{\cos^2 x} + \frac{\cos^2 x}{\cos^2 x} = \frac{1}{\cos^2 x} \Leftrightarrow \tan^2 x + 1 = \sec^2 x$$

So $\displaystyle\int \tan^2 x\,dx = \int (\sec^2 x - 1)\,dx = \tan x - x + c$

EXERCISE 2A

Find the following indefinite and definite integrals.

1 $\int (e^{2x} + x) \, dx$

2 $\int \frac{1}{3x} \, dx$

3 $\int \frac{1}{3x + 2} \, dx$

4 $\int \frac{1}{\cos^2 x} \, dx$

5 $\int_0^{\frac{\pi}{6}} (\sec^2 x - \sin x) \, dx$

6 $\int_0^1 e^{x+2} \, dx$

7 $\int_{-1}^1 (e^x - 1)^2 \, dx$

8 $\int_2^1 \left(\frac{2}{x} + \frac{x}{2} \right) dx$

EXERCISE 2B

Find the following definite integrals.

1 $\int_0^{\frac{\pi}{4}} \left(\frac{1 + \sin^2 x}{\cos^2 x} \right) dx$

2 $\int_2^3 \left(\frac{1}{x - 1} - \frac{1}{x + 1} \right) dx$

3 $\int_{\frac{\pi}{6}}^{\frac{\pi}{4}} (\sin^3 x \, \text{cosec}^2 x) \, dx$

4 $\int_0^1 \frac{(e^x - 1)^2}{e^{2x}} \, dx$

5 $\int_2^3 \left(\frac{1}{2e^{3x}} \right) dx$

6 $\int_e^{2e} \frac{4}{x} \, dx$

7 $\int_{-\frac{\pi}{2}}^{\frac{\pi}{2}} \left(\frac{\cos x}{\cos^2 \frac{x}{2}} \right) dx$

8 $\int_0^1 \left(\frac{x - 3}{x^2 - 9} \right) dx$

Differials

If $y = f(x)$ and if δy is the small change in y induced by a small change δx in x, then

$$\delta y \approx \lim_{\delta x \to 0} \frac{\delta y}{\delta x}.\delta x$$

i.e. $\delta y \approx \frac{dy}{dx}.\delta x$

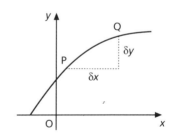

The smaller δx becomes, the better is the approximation for δy.

With this idea in mind, we define two quantities dy, the y-differential, and dx, the x-differential, by the equation $dy = \frac{dy}{dx} \, dx$.

Note that $\frac{dy}{dx}$ is the coefficient of dx and is often referred to as the *differential coefficient*.

Now, if for example $y = x^2$, then we can write $dy = 2x \, dx$.

Integration by substitution

When differentiating a composite function we made use of the chain rule. For example, to differentiate $y = (2x + 3)^3$:

Let $u = 2x + 3$. Then $y = u^3 \Rightarrow \dfrac{dy}{du} = 3u^2$ and $\dfrac{du}{dx} = 2$

$$\frac{dy}{dx} = \frac{dy}{du}.\frac{du}{dx} = 3u^2.2 = 6u^2$$

$$\Rightarrow \frac{dy}{dx} = 6(2x + 3)^2$$

When integrating we must reduce the function to a *standard form*: one for which we know an antiderivative. This can be awkward but, under certain conditions, we can use the chain rule *in reverse*.

If we wish to perform an integral of the form

$$\int g(f(x)).\, f'(x)\, dx$$

we can proceed as follows.

Let $u = f(x)$. Then, using differentials, $du = f'(x)\, dx$ and the integration takes the form

$$\int g(u)\, du \quad \text{which it is hoped will be a standard form}$$

For reference we will call $f(x)$ the *essential* function.

Example 1   Find $\displaystyle\int x(x^2 + 3)^3\, dx$.

The essential function is $f(x) = x^2 + 3$.
Let $u = x^2 + 3$ then $du = 2x\, dx$ and so $x\, dx = \frac{1}{2} du$.
Substituting, the integral becomes:

$$\int \frac{1}{2} u^3 \, du = \frac{1}{8} u^4 + c$$

And substituting back:

$$\int x(x^2 + 3)^3 \, dx = \frac{1}{8}(x^2 + 3)^4 + c$$

Example 2   Find $\displaystyle\int 3x^2(x^3 - 4)^5\, dx$.

Let $u = x^3 - 4$. Then $du = 3x^2\, dx$.
Substituting gives:

$$\int u^5 \, du = \frac{1}{6} u^6 + c$$

Substituting back:

$$\int 3x^2(x^3 - 4)^5 \, dx = \frac{1}{6}(x^3 - 4)^6 + c$$

Example 3   Find $\int 8 \cos x \sin^3 x \, dx$.

Let $u = \sin x$. Then $du = \cos x \, dx$.
Substituting gives:
$$\int 8u^3 \, du = 2u^4 + c$$
Substituting back:
$$\int 8 \cos x \sin^3 x \, dx = 2 \sin^4 x + c$$

EXERCISE 3

Find the following indefinite integrals.
In each case identify $f(x)$, the essential function, and then make the substitution $u = f(x)$.

1 $\int 2x(x^2 + 3)^7 \, dx$
2 $\int 5x^2(x^3 - 4)^4 \, dx$
3 $\int (2x + 3)(x^2 + 3x)^5 \, dx$

4 $\int (2 - x)(4x - x^2)^3 \, dx$
5 $\int (3x^2 - 5x)^5(6x - 5) \, dx$
6 $\int 3(3x^2 + 4x)(x^3 + 2x^2)^4 \, dx$

7 $\int e^x \sin(e^x) \, dx$
8 $\int 2x \, e^{x^2} \, dx$
9 $\int x^2 e^{x^3 + 3} \, dx$

10 $\int \cos x \, e^{\sin x} \, dx$
11 $\int x(2x^2 + 1)^3 \, dx$
12 $\int x^2(4x^3 - 2)^{-2} \, dx$

13 $\int 3x(x^2 + 2)^6 \, dx$
14 $\int \cos x \sin^4 x \, dx$ [Hint: essential function is $\sin x$.]

15 $\int 4 \sin x \cos^3 x \, dx$
16 $\int \sec^2 x \tan^4 x \, dx$ [Hint: essential function is $\tan x$.]

17 $\int 4 \, \text{cosec}^2 x \cot^4 x \, dx$ [Hint: what is the derivative of $\cot x$?]

18 $\int \sec^3 x \tan x \, dx$ [Hint: write as $\sec x \tan x \, (\sec^2 x)$.]

19 $\int \text{cosec}^4 x \cot x \, dx$

For many questions, the choice of essential function is not obvious.
In these cases one will be suggested. Often there is a variety of ways at arriving at the same conclusion.

Example 1   Find $= \int \dfrac{4 \ln x}{x} \, dx$ (let $u = \ln x$).

$$du = \frac{1}{x} dx$$

Substituting gives:
$$\int 4u \, du = 2u^2 + c$$

Substituting back:
$$\int \frac{4 \ln x}{x} \, dx = 2(\ln x)^2 + c$$

74

Example 2 Find $= \int \sqrt{(1 - x^2)}\, dx$ (let $u = \sin^{-1} x$).

$x = \sin u$

$\Rightarrow dx = \cos u\, du$

Substituting gives:

$$\int \sqrt{(1 - \sin^2 u)} \cos u\, du$$

$$= \int \cos^2 u\, du = \int \frac{1}{2} \cos 2u - \frac{1}{2}\, du \qquad \text{using } \cos 2u = 2 \cos^2 u - 1$$

$$= \frac{1}{4} \sin 2u - \frac{1}{2} u + c$$

$$= \frac{1}{4} 2 \sin u \cos u - \frac{1}{2} u$$

$$= \frac{1}{2} \sin u \sqrt{(1 - \sin u)} - \frac{1}{2} u + c$$

Substituting back:

$$\int \sqrt{(1 - x^2)}\, dx = \frac{1}{2} x \sqrt{(1 - x^2)} - \frac{1}{2} \sin^{-1} x + c$$

Some trigonometric reminders

$\sin^2 x + \cos^2 x = 1$

$\sin 2x = 2 \sin x \cos x$

$\cos 2x = \cos^2 x - \sin^2 x$

$\qquad = 2 \cos^2 x - 1$

$\qquad = 1 - 2 \sin^2 x$

$\cos^2 x = \frac{1}{2}(1 + \cos 2x)$

$\sin^2 x = \frac{1}{2}(1 - \cos 2x)$

EXERCISE 4A

Use the suggested substitution to help you find the following integrals.

1 $\int (2x + 4)\sqrt{x^2 + 4x}\, dx$ (let $u = x^2 + 4x$)

2 $\int \dfrac{x\, dx}{(x^2 - 1)^{\frac{5}{2}}}$ (let $u = x^2 - 1$)

3 $\int (3x^2 - 2)\sqrt{x^3 - 2x + 1}\, dx$ (let $u = x^3 - 2x + 1$)

4 $\int 5 \sec^2 x \tan^4 x\, dx$ (let $u = \tan x$)

5 $\int \dfrac{(\ln x)^3}{x}\, dx$ (let $u = \ln x$)

6 $\int x\, e^{3x^2 + 2}\, dx$ (let $u = 3x^2 + 2$)

7 $\int \dfrac{e^{\sqrt{x}}}{\sqrt{x}}\, dx$ (let $u = \sqrt{x}$)

8 $\int \tan x\, dx$ (let $u = \cos x$)

9 $\int x\sqrt{a^2 + x^2}\, dx$ (let $u^2 = a^2 + x^2$)

10 $\int \sqrt{1 - x^2}\, dx$ (let $x = \sin u$)

11 $\int \sin x \cos x \cos^3 2x\, dx$ (let $\cos 2x = u$)

12 $\int \dfrac{x + 3}{\sqrt{(x - 1)}}\, dx$ (let $u = x - 1$)

Example 1   Find $\int \sin^5 x \, dx$.

$$\int \sin^5 x \, dx = \int \sin^4 x \sin x \, dx = \int (\sin^2 x)^2 \sin x \, dx = \int (1 - \cos^2 x)^2 \sin x \, dx$$

Let $u = \cos x$. Then $du = -\sin x \, dx$.

Substituting gives:

$$\int -(1 - u^2)^2 \, du = \int -(1 - 2u^2 + u^4) \, du = -u + \frac{2}{3}u^3 - \frac{1}{5}u^5 + c$$

Substituting back:

$$\int \sin^5 x \, dx = -\cos x + \frac{2}{3}\cos^3 x - \frac{1}{5}\cos^5 x + c$$

Sometimes it is simpler finding the differential dx in terms of u.

Example 2   Find $\int \dfrac{1}{1 - \sqrt{x}} \, dx \ (x \neq 1)$.

Let $u = 1 - \sqrt{x}$. Then $du = -\frac{1}{2}x^{-\frac{1}{2}} dx \Rightarrow dx = -2x^{\frac{1}{2}} du \Rightarrow dx = -2(1 - u) \, du$

Substituting gives:

$$\int \frac{-2(1 - u)}{u} \, du = \int (-2u^{-1} + 2) \, du = -2 \ln|u| + 2u + c$$

Substituting back:

$$\int \frac{1}{1 - \sqrt{x}} \, dx = -2 \ln|1 - \sqrt{x}| + 2(1 - \sqrt{x}) + c$$

EXERCISE 4B

Work out the following integrals. Hints have sometimes been provided.

1 $\int \sin^3 x \, dx$ **2** $\int \cos^3 x \, dx$ (let $u = \sin x$)   **3** $\int \cos^5 x \, dx$

4 $\int \cos^4 x \, dx$ [Hint: $\cos^4 x = \cos^2 x . \cos^2 x$ *and* $\cos^2 x = (1 + \cos 2x)/2$.]

5 $\int \sin^4 x \, dx$ **6** $\int \sec^4 x \, dx$ [Hint: $\tan^2 x + 1 = \sec^2 x$.]

7 $\int \dfrac{dx}{1 + \sqrt{x}}$ (let $u = 1 + \sqrt{x}$)   **8** $\int \dfrac{x \, dx}{\sqrt{2 + 4x}}$ (let $u^2 = 2 + 4x$)

9 $\int \dfrac{x \, dx}{(a + bx)^{\frac{3}{2}}}$ (let $u^2 = a + bx$)   **10** $\int \dfrac{dx}{x^4(1 + x^2)^{\frac{1}{2}}}$ $\left(\text{let } x = \dfrac{1}{\sqrt{(u^2 - 1)}}\right)$

11 $\int \dfrac{x^3 dx}{(1 + x^2)^{\frac{3}{2}}}$ (let $x = \sqrt{(u^2 - 1)}$)   **12** $\int x\sqrt[3]{1 + x} \, dx$ (let $u^3 = 1 + x$)

13 $\int \dfrac{dx}{x^2\sqrt{x^2 + 1}}$ (let $x = \tan u$)

14 $\displaystyle\int \dfrac{dx}{\sin x}$ $\begin{cases} \text{• express } \sin x \text{ as } 2\sin\dfrac{x}{2}\cos\dfrac{x}{2} \\[2mm] \text{• multiply numerator and denominator by } \sec^2\dfrac{x}{2} \\[2mm] \text{• let } u = \tan\dfrac{x}{2} \end{cases}$

Substitution and definite integrals

Assuming the function is continuous over the interval of integration, then exchanging the limits for x by the corresponding limits for u will save you having to substitute back after the integration process.

Example   Evaluate $\displaystyle\int_1^2 (2x+4)(x^2+4x)^3\,dx$.

Let $u = x^2 + 4x$. Then $du = (2x+4)\,dx$; also, when $x = 2$, $u = 12$; when $x = 1$, $u = 5$.
Substituting gives:

$$\int_5^{12} u^3\,du = \left[\dfrac{u^4}{4}\right]_5^{12} = 5027.75$$

EXERCISE 5A

Find the following definte integrals using the given substitution.

1 $\displaystyle\int_0^1 x^3(2+x^4)^5\,dx,\ u = x^4 + 2$

2 $\displaystyle\int_4^9 \dfrac{x+1}{x^2+2x-7}\,dx,\ u = x^2 + 2x - 7$

3 $\displaystyle\int_0^4 \dfrac{\sqrt{x}}{2+\sqrt{x}}\,dx,\ u = \sqrt{x} + 2$

4 $\displaystyle\int_0^{\frac{1}{2}} \dfrac{x}{\sqrt{1-x^2}}\,dx,\ u = 1 - x^2$

5 $\displaystyle\int_0^1 x\sqrt{2x^2+3}\,dx,\ u = 2x^2 + 3$

6 $\displaystyle\int_{\frac{\pi}{3}}^{\frac{\pi}{2}} \dfrac{\sin x}{1+\cos x}\,dx,\ u = 1 + \cos x$

7 $\displaystyle\int_0^{\frac{\pi}{2}} \dfrac{1-2\sin x}{x+2\cos x}\,dx,\ u = x + 2\cos x$

8 $\displaystyle\int_{\frac{\pi}{3}}^{\frac{\pi}{2}} \cot x\,dx,\ u = \sin x$

9 $\displaystyle\int_0^{\frac{\pi}{3}} \sin x\cos^4 x\,dx,\ u = \cos x$

10 $\displaystyle\int_{\frac{\pi}{6}}^{\frac{\pi}{4}} \dfrac{\sin x}{\cos^2 x}\,dx,\ u = \cos x$

11 $\displaystyle\int_e^{e^2} \dfrac{1}{x\ln x}\,dx,\ t = \ln x$

12 $\displaystyle\int_1^2 \dfrac{e^{2x}}{(e^{2x}-1)^2}\,dx,\ t = e^{2x} - 1$

EXERCISE 5B

1 Find $\displaystyle\int_0^{\frac{\pi}{2}} \sin^5 x\,dx$, using the substitution $u = \cos x$ and remembering that
$\sin^2 x + \cos^2 x = 1$.

2 Find $\displaystyle\int_0^{\frac{\pi}{4}} \sin^3 x\cos^2 x\,dx$ making the substitution $u = \cos x$.

3 a Show that $\dfrac{8}{u^2 - 4}$ can be written as $\dfrac{2}{u - 2} - \dfrac{2}{u + 2}$.

  b Hence, by making the substitution $u = \sqrt{x + 4}$, find $\displaystyle\int_5^{12} \dfrac{\sqrt{x + 4}}{x}\,dx$.

4 a Show that $\dfrac{1}{u^2 - 1} = \dfrac{\frac{1}{2}}{u - 1} - \dfrac{\frac{1}{2}}{u + 1}$.

  b Hence, by making the substitution $u = \sqrt{1 + x^2}$, find $\displaystyle\int_{\frac{3}{4}}^{\frac{4}{3}} \dfrac{\sqrt{1 + x^2}}{x^3 + x}\,dx$.

5 Find $\displaystyle\int_{\frac{1}{2}}^{1} \dfrac{1 - x}{\sqrt{1 - x^2}}\,dx$ making the substitution $x = \sin u$.

6 Find $\displaystyle\int_{\frac{\pi}{4}}^{\frac{\pi}{3}} \dfrac{1}{\sin^2 x \sqrt{3 - \cot x}}\,dx$. Here you will need to decide which is the essential function.

Discontinuities

If a function, f, is defined on an interval which has one end open at $x = a$ then, under certain circumstances, the area under the curve up to $x = a$ can still be evaluated. The two examples below should illustrate this.

Example 1   Consider $\displaystyle\int_0^1 \dfrac{2x\,dx}{(x^2 - 1)^{\frac{2}{3}}}$.

As shown in the sketch, the function is undefined at $x = 1$.

Let us, however, consider $\displaystyle\int_0^{1-\varepsilon} \dfrac{2x\,dx}{(x^2 - 1)^{\frac{2}{3}}}$ and examine its limit as ε tends to zero.

$$\lim_{\varepsilon \to 0} \int_0^{1-\varepsilon} \dfrac{2x\,dx}{(x^2 - 1)^{\frac{2}{3}}} = \lim_{\varepsilon \to 0}\left[3(x^2 - 1)^{\frac{1}{3}} \right]_0^{1-\varepsilon}$$

$$= \lim_{\varepsilon \to 0}\left[3((1 - \varepsilon)^2 - 1)^{\frac{1}{3}} - 3(-1)^{\frac{1}{3}} \right]$$

$$= 3 \times 0 - 3 \times (-1)$$

$$= 3$$

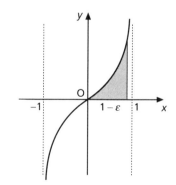

The limit exists and we write $\displaystyle\int_0^1 \dfrac{2x\,dx}{(x^2 - 1)^{\frac{2}{3}}} = 3$

Example 2   Consider the function $f(x) = \dfrac{1}{(x-1)^2}$.

This function is undefined at $x = 1$. If we wish to investigate the area under the curve from 0 to 2 then we would have to consider integrations in the intervals $[0, 1)$ and $(1, 2]$.

$$\lim_{\varepsilon \to 0} \int_0^{1-\varepsilon} \frac{dx}{(x-1)^2} = \lim_{\varepsilon \to 0}\left(\frac{1}{\varepsilon} - 1\right)$$

and $\displaystyle \lim_{\varepsilon \to 0} \int_{1+\varepsilon}^{2} \frac{dx}{(x-1)^2} = \lim_{\varepsilon \to 0}\left(-1 + \frac{1}{\varepsilon}\right)$

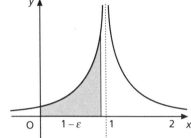

Both limits tend to infinity as ε tends to zero: the area is unbounded.

If the area under a curve is required and the function is undefined at some point in the desired interval, two integrals up to the point of discontinuity have to be examined separately. Proceed with caution.

Similar caution should be extended to finding the area under a piecewise function, f, when a point where $f'(x)$ is undefined lies within the desired interval.

Special (common) forms

Some substitutions are so common that they can be treated as *standards* and, when their form is established, their integrals can be written down without further ado.

1 $\displaystyle \int f(ax + b)\, dx$ Let $u = ax + b$. Then $du = a\, dx \Rightarrow dx = \dfrac{du}{a}$.

Substituting gives:

$$\frac{1}{a}\int f(u)\, du = \frac{1}{a}F(u) + c \text{ where } F \text{ is an antiderivative of } f$$

Substituting back:

$$\int f(ax + b)\, dx = \frac{1}{a}F(ax + b) + c$$

Example 1 $\displaystyle \int (3x + 2)^4\, dx = \frac{1}{3}\cdot\frac{1}{5}(3x + 2)^5 + c$

Example 2 $\displaystyle \int \cos(2 - 3x)\, dx = -\frac{1}{3}\sin(2 - 3x) + c$

Example 3 $\displaystyle \int e^{(2 - 3x)}\, dx = -\frac{1}{3}e^{(2 - 3x)} + c$

2 $\displaystyle\int \frac{f'(x)}{f(x)}\,dx$ Let $u = f(x)$. Then $du = f'(x)\,dx$.

Substituting gives:

$$\int \frac{du}{u} = \ln|u| + c$$

Substituting back:

$$\int \frac{f'(x)}{f(x)}\,dx = \ln|f(x)| + c$$

Example 1 $\displaystyle\int \frac{2x + 3}{x^2 + 3x}\,dx = \ln|x^2 + 3x| + c$

Example 2 $\displaystyle\int \frac{x}{x^2 + 4}\,dx = \frac{1}{2}\int \frac{2x}{x^2 + 4}\,dx = \frac{1}{2}\ln|x^2 + 4| + c$

Example 3 $\displaystyle\int \cot x\,dx = \int \frac{\cos x}{\sin x}\,dx = \ln|\sin x| + c$

3 $\displaystyle\int f'(x)f(x)\,dx$ Let $u = f(x)$. Then $du = f'(x)\,dx$.

Substituting gives:

$$\int u\,du = \frac{1}{2}u^2 + c$$

Substituting back:

$$\int f'(x)f(x)\,dx = \frac{1}{2}(f(x))^2 + c$$

Example 1 $\displaystyle\int (2x + 3)(x^2 + 3x)\,dx = \frac{1}{2}(x^2 + 3x)^2 + c$

Example 2 $\displaystyle\int x(x^2 + 4)\,dx = \frac{1}{2}\int 2x(x^2 + 4)\,dx = \frac{1}{2}\cdot\frac{1}{2}(x^2 + 4)^2 + c$

EXERCISE 6A

Find the following integrals.

1 a $\displaystyle\int \sqrt{2 - x}\,dx$ **b** $\displaystyle\int (x^2 + 2)^2\,dx$ **c** $\displaystyle\int \frac{2x}{2 + x^2}\,dx$ **d** $\displaystyle\int \frac{e^{2x}}{1 + e^{2x}}\,dx$

Find the following integrals using the given substitution.

2 $\displaystyle\int_{e}^{e^2} \frac{(1 + \ln x)^2}{x}\,dx$; let $1 + \ln x = t$ **3** $\displaystyle\int \frac{\cos \sqrt{x}}{\sqrt{x}}\,dx$; let $x = t^2$

4 $\displaystyle\int \frac{1}{x(1 + \ln x)^3}\,dx$; let $1 + \ln x = t$ **5** $\displaystyle\int_{0}^{3} \frac{x^2}{\sqrt{1 + x}}\,dx$; let $1 + x = t^2$

6 $\displaystyle\int \frac{1}{1 + \sqrt{x}}\,dx$; let $1 + \sqrt{x} = t$ **7** $\displaystyle\int_{0}^{1} x^2(x^3 - 2)^4\,dx$; let $x^3 - 2 = t$

8 $\displaystyle\int \frac{\ln x}{x}\,dx$; let $\ln x = t$ **9** $\displaystyle\int \cos x(1 - \sin x)^n\,dx$; let $1 - \sin x = t$

10 $\displaystyle\int_0^{\frac{\pi}{4}} \frac{\sec^2 x}{1 + \tan x}\,dx$; let $1 + \tan x = u$

11 $\displaystyle\int \frac{4x + 5}{\sqrt{2x - 3}}\,dx$; let $2x - 3 = t^2$

12 $\displaystyle\int \frac{\sqrt{x}}{\sqrt{x} - 2}\,dx$; let $\sqrt{x} - 2 = t$

13 $\displaystyle\int_0^{\frac{\pi}{3}} \cos^2 x \sin^3 x\,dx$; let $\cos x = t$

14 $\displaystyle\int \tan x \sec^3 x\,dx$; let $\cos x = t$

15 $\displaystyle\int \sin^5 x\,dx$; let $\cos x = t$

EXERCISE 6B

1 Show that $1 + \dfrac{4}{x - 2} = \dfrac{x + 2}{x - 2}$ and hence find $\displaystyle\int \frac{x + 2}{x - 2}\,dx$.

2 Find $\displaystyle\int \frac{\sin x}{(a - b \cos x)^4}\,dx$ using the substitution $a - b \cos x = t$.

3 a Show that $\dfrac{e^x}{1 + e^x} - \dfrac{-e^x}{1 - e^x} = \dfrac{2e^x}{1 - e^{2x}}$.

  b Hence show that $\displaystyle\int \frac{e^x}{1 - e^{2x}}\,dx = \ln\sqrt{\frac{1 + e^x}{1 - e^x}}$.

4 Find $\displaystyle\int \tan^n x \sec^2 x\,dx$ using the substitution $\tan x = v$.

5 Find $\displaystyle\int_1^e \frac{1}{x}(\ln x)^n\,dx$ using the substitution $\ln x = v$.

6 a Show that $\dfrac{\sec^2 x}{1 + \tan x} + \dfrac{\sec^2 x}{1 - \tan x} = \dfrac{2\sec^2 x}{1 - \tan^2 x}$.

  b Hence find $\displaystyle\int \frac{\sec^2 x}{1 - \tan^2 x}\,dx$.

7 Find $\displaystyle\int \frac{1}{\sqrt{2x + 7} + 12}\,dx$ using the substitution $2x + 7 = (t - 12)^2$.

8 Find $\displaystyle\int \frac{\sin 2x}{3 + 4\cos^2 x}\,dx$ using the substitution $3 + 4\cos^2 x = t$.

9 Using the substitution $x = \tan t$, show that $\displaystyle\int_0^1 \frac{x - 3}{(1 + x^2)^{\frac{3}{2}}}\,dx = 2\sqrt{2} - 1$.

10 a Find $\displaystyle\int_3^4 \frac{1}{(x - 1)^3}\,dx$.

  b The function $f(x) = \dfrac{1}{(x - 1)^3}$ has a discontinuity at $x = 1$ so $\displaystyle\int_0^2 \frac{1}{(x - 1)^3}$

  is not necessarily meaningful. By separating the integral into

  $\displaystyle\int_0^{1-\varepsilon} \frac{1}{(x - 1)^3}\,dx$ and $\displaystyle\int_{1+\varepsilon}^2 \frac{1}{(x - 1)^3}\,dx$ and examining the limits as $\varepsilon \to 0$ decide

  whether the integral is meaningful and, if so, what its value is.

11 Show that $\displaystyle\int_0^2 \frac{3x^2}{(x^3 - 1)^{\frac{1}{3}}}\,dx$ is a meaningful integral and find its value.

Area under a curve

Early estimates of the area under a curve were rather crude and inaccurate but they did provide an estimate!

To illustrate the method, look at the area shaded under the curve $y = x$ between $x = 1$ and $x = 2$.

For a first estimate, divide the 'base' into five equal parts and draw rectangles as shown. The heights of the rectangles and the corresponding areas are calculated (using a spreadsheet, for example) as

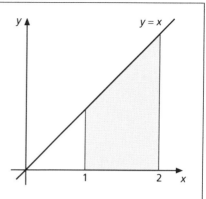

x	1	1.2	1.4	1.6	1.8
Height	1	1.2	1.4	1.6	1.8
Area	0.2	0.24	0.28	0.32	0.36

The total area of the five rectangles is 1.4 and since we know what the actual area is (here it is just a trapezium with area = 1.5) we can see we have a percentage error of about 6%.

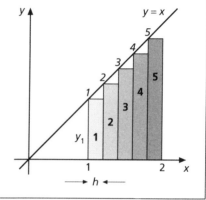

EXERCISE 7

1 Using a similar approach find an estimate for the area under the curve $y = x$ between $x = 1$ and $x = 2$ using 10 rectangles.

2 Using a similar approach find estimates for the area under the curve $y = x^2$ between $x = 1$ and $x = 2$ using **(i)** 5 rectangles   **(ii)** 10 rectangles.

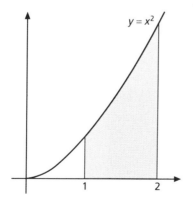

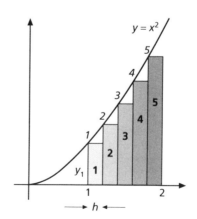

Clearly, the more rectangles we can use, the closer the estimate gets to the actual area. So the next step is to divide the 'base' into n equal parts, find an expression for the sum of the areas of these n rectangles and then examine whether there is a limit as $n \to \infty$.

This process could be called 'Integration from first principles' and, on the face of it, it looks fairly sound. Unfortunately, unlike its companion (differentiation from first principles), it is anything but. As we shall see from the next illustration, it depends on knowing formulae for the sums of series and only with the simplest of functions is this practical. The process is illustrated with the area under the simplest of curves, namely $y = x$.

Example By summing the areas of rectangles, show that the area under the curve $y = x$ between $x = a$ and $x = b$ is given by $\frac{1}{2}(b^2 - a^2)$.

Divide the 'base' $b - a$ into n intervals, each of width h, so that $\frac{b - a}{n} = h$.
A table of values is

Rectangle	1	2	3	4	5	$\cdots$	n
x	a	$a + h$	$a + 2h$	$a + 3h$	$a + 4h$	$\cdots$	$a + (n - 1)h$
Height	a	$a + h$	$a + 2h$	$a + 3h$	$a + 4h$	$\cdots$	$a + (n - 1)h$
Area	ha	$h(a + h)$	$h(a + 2h)$	$h(a + 3h)$	$h(a + 4h)$	$\cdots$	$h(a + (n - 1)h)$

The sum of the areas of the rectangles is

$ha + h(a + h) + h(a + 2h) + h(a + 3h) + h(a + 4h) + \cdots + h(a + (n - 1)h)$

$= n \times ha + h^2 + 2h^2 + 3h^3 + \cdots + (n - 1)h^2$

$= nha + h^2(1 + 2 + 3 + 4 + \cdots + (n - 1))$

$= nha + \frac{1}{2}h^2.(n - 1).n$

$= nha + \frac{1}{2}n^2h^2 - \frac{1}{2}nh^2$

$= (b - a)a + \frac{1}{2}(b - a)^2 - \frac{1}{2}h(b - a)$

> We need to know that
> $1 + 2 + 3 + 4 + \cdots + n = \frac{1}{2}n(n + 1)$

> Since $\frac{b - a}{n} = h$ then $nh = b - a$

As we require the limit as $n \to \infty$, which is equivalent to $h \to 0$, the aim is to end with an expression involving only one of these, say h, and to ensure there are no hs in the bottom of fractions.

The area under the curve $y = x$ equals $\quad \lim_{h \to 0}\left[(b - a)a + \frac{1}{2}(b - a)^2 - \frac{1}{2}h(b - a)\right]$

$$= (b - a)a + \frac{1}{2}(b - a)^2$$

$$= ba - a^2 + \frac{1}{2}b^2 - ab + \frac{1}{2}a^2$$

$$= \frac{1}{2}b^2 - \frac{1}{2}a^2$$

EXERCISE 8

Using a similar process show that area under the curve $y = x^2$ from $x = a$ to $x = b$ is $\frac{1}{3}b^3 - \frac{1}{3}a^3$. You will need the following formulae:

$$1 + 2 + 3 + 4 + \cdots + n = \frac{1}{2}n(n + 1)$$

and

$$1^2 + 2^2 + 3^2 + 4^2 + \cdots + n^2 = \frac{1}{6}n(n + 1)(2n + 1)$$

You should have:

sum of rectangles

$= h \times na^2 + h \times 2ah(1 + 2 + 3 + \cdots + (n - 2)) + h \times h^2(1^2 + 2^2 + 3^2 + \cdots + (n - 2)^2)$

$= hna^2 + 2ah^2(\frac{1}{2}(n - 2)(n - 1)) + h^3(\frac{1}{6}(n - 2)(n - 1)(2n - 3))$

$= hna^2 + ah^2n^2 - 3anh^2 + 2ah^2 + \frac{1}{3}h^3n^3 - \frac{3}{2}h^3n^2 + \frac{13}{6}h^3n - h$

$= (b - a)a^2 + a(b - a)^2 - h[3a(b - a) + 2ah] + \frac{1}{3}(b - a)^3 + h\left[-\frac{3}{2}(b - a)^2 + \frac{13}{6}h(b - a) - h^2\right]$

$\lim_{h \to 0}(\ldots) = (b - a)a^2 + a(b - a)^2 + \frac{1}{3}(b - a)^3 = \ldots \frac{1}{3}(b^3 - a^3)$

Applications
Areas between the curve and the x-axis

Reminders

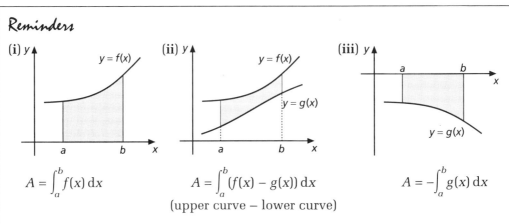

$$A = \int_a^b f(x)\,dx \qquad\qquad A = \int_a^b (f(x) - g(x))\,dx \qquad\qquad A = -\int_a^b g(x)\,dx$$

(upper curve – lower curve)

All areas can be thought of as 'upper curve – lower curve'.

In **(i)** the 'lower curve' is the x-axis, i.e. the function $f(x) = 0$.

so $A = \int_a^b (f(x) - g(x))\,dx$

$= \int_a^b (f(x) - 0)\,dx$

$= \int_a^b f(x)\,dx$

In **(iii)** the 'upper curve' is the x-axis, i.e. the function $g(x) = 0$.

so $A = \int_a^b (f(x) - g(x))\,dx$

$= \int_a^b (0 - g(x))\,dx$

$= -\int_a^b g(x)\,dx$

Areas enclosed or partly enclosed need careful treatment. In this diagram the area marked A_1 has a different upper curve to the part marked A_2 so the two parts needed to be treated separately.

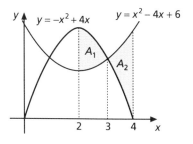

$$A_1 = \int_2^3 ((-x^2 + 4x) - (x^2 + 3x - 15))\,dx$$

$$A_2 = \int_3^4 ((x^2 + 3x - 15) - (-x^2 + 4x))\,dx$$

$$A = A_1 + A_2 = \ldots = 4\tfrac{5}{6} + 6\tfrac{1}{6} = 11.$$

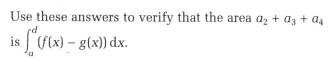

Note

If the value of x at the intersection (i.e. $x = 3$) is not given, then you would have to calculate this first.

EXERCISE 9

In this diagram the shaded part has the same 'upper curve' all the time so the area is
$\int_a^d (f(x) - g(x))\,dx$, even though there are parts above and below the x-axis. To justify this write down the integrals for each of the areas a_1 to a_6. (Note that some of the areas require more than one integral.)

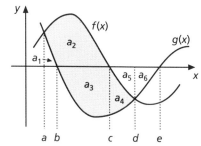

Use these answers to verify that the area $a_2 + a_3 + a_4$ is $\int_a^d (f(x) - g(x))\,dx$.

You should have the following:

$$a_1 = \int_a^b g\,dx, \quad a_1 + a_2 = \int_a^c f\,dx \text{ so } a_2 = \int_a^c f\,dx - \int_a^b g\,dx, \quad a_3 = -\int_b^c g\,dx$$

$$a_4 + a_5 = -\int_c^d g\,dx, \quad a_5 = -\int_c^d f\,dx \text{ so } a_4 = -\int_c^d g\,dx + \int_c^d f\,dx, \quad a_6 = -\int_d^e g\,dx$$

Therefore
$$a_2 + a_3 + a_4 = \int_a^c f\,dx - \int_a^b g\,dx - \int_b^c g\,dx - \int_c^d g\,dx + \int_c^d f\,dx$$

$$= \int_a^c f\,dx + \int_c^d f\,dx - \left(\int_a^b g\,dx + \int_b^c g\,dx + \int_c^d g\,dx \right)$$

$$= \int_a^d f\,dx - \left(\int_a^e g\,dx \right)$$

$$= \int_a^d (f - g)\,dx$$

Areas between the curve and the y-axis

Just as the area between a function and the x-axis (dotted) is found by calculating $\int_a^b f(x)\,dx$, in an analogous situation the area between the function and the y-axis (striped) can be found by calculating $\int_c^d f^{-1}(y)\,dy$.

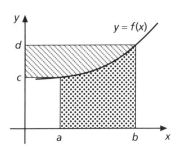

Care must be exercised. First of all, this can only be done when x can be expressed as a function of y (i.e. *can* we write $x = f^{-1}(y)$?) – this is not always possible.

Secondly, it may be quicker to find the area represented by $\int_a^b f(x)\,dx$ and subtract it from an appropriate rectangular shape.

Note that, if you use $\int f^{-1}(y)\,dy$ to find the area between two curves, then the area is given by $\int_a^b (\text{'right-hand' curve} - \text{'left-hand' curve})\,dy$.

Example   Calculate the striped shaded area in the diagram.

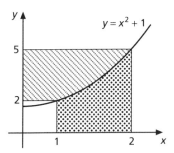

Method 1

Dotted area $= \displaystyle\int_1^2 (x^2 + 1)\,dx$

$\qquad\qquad = \left[\dfrac{1}{3}x^3 + x\right]_1^2$

$\qquad\qquad = \left(\dfrac{8}{3} + 2\right) - \left(\dfrac{1}{3} + 1\right)$

$\qquad\qquad = \dfrac{10}{3}$

Striped area $= 10 - 2 - \dfrac{10}{3}$

$\qquad\qquad = \dfrac{14}{3}$

Method 2

Since
$y = x^2 + 1$ then $x = \sqrt{y - 1}$

so striped area

$\qquad = \displaystyle\int_2^5 \sqrt{y - 1}\,dy$

$\qquad = \left[\dfrac{2}{3}(y - 1)^{\frac{3}{2}}\right]_2^5$

$\qquad = \left(\dfrac{2}{3} \times 4^{\frac{3}{2}}\right) - \left(\dfrac{2}{3} \times 1^{\frac{3}{2}}\right)$

$\qquad = \dfrac{14}{3}$

Volumes of revolution

Volumes of revolution are formed when a curve is rotated about the x-axis (or the y-axis). A simple case is when the line $y = kx$ is rotated about the x-axis – this forms a cone. The formula for such volumes can be arrived at by using a similar procedure to slicing the area up into strips as in Exercise 7. For volumes the slices become thin discs and the procedure becomes very difficult for anything other than simple functions!

The end product is that

$V = \int_a^b \pi y^2 \, dx$ for   and $V = \int_c^d \pi x^2 \, dy$ for

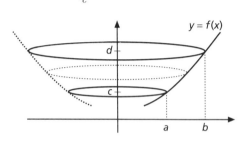

Example Find the volume of revolution obtained between $x = 1$ and $x = 2$ when the curve $y = x^2 + 2$ is rotated

(i) about the x-axis

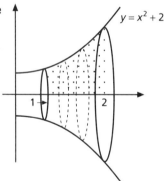

(ii) about the y-axis.

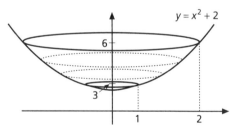

(i) $V = \int_1^2 \pi y^2 \, dx$

$= \pi \int_1^2 (x^4 + 4x^2 + 1) \, dx$

$= \pi \left[\frac{1}{5}x^5 + \frac{4}{3}x^3 + x \right]_1^2$

$= \pi \left(\frac{32}{5} + \frac{32}{3} + 2 \right) - \pi \left(\frac{1}{5} + \frac{1}{3} + 1 \right)$

$= \frac{263}{15}\pi$

(ii) When $x = 1$, $y = 3$ and, when $x = 2$, $y = 6$

$V = \int_3^6 \pi x^2 \, dy$

$= \pi \int_3^6 (y - 2) \, dy$

$= \pi \left[\frac{1}{2}y^2 - 2y \right]_3^6$

$= \pi (18 - 12) - \pi \left(4\frac{1}{2} - 6 \right)$

$= \frac{15}{2}\pi$

Displacement, velocity, acceleration for rectilinear motion

We have already seen in Chapter 2 on differential calculus that, where the distance is given as a function of time, i.e. $s = f(t)$, then the velocity v is given by $v = \dfrac{ds}{dt}$ and the acceleration a by $a = \dfrac{d}{dt}(v) = \dfrac{d}{dt}\left(\dfrac{ds}{dt}\right) = \dfrac{d^2s}{dt^2}$. Using antiderivatives we can now 'reverse' these rules so that, given any one of distance, velocity and acceleration, and appropriate values, we can determine the remaining two.

> **Note**
> In the following context, 'distance' is used to mean displacement from the origin.

Example A particle starts from rest and, at time t seconds, the velocity is given by $v = 3t^2 + 4t - 1$. Determine the distance, velocity and acceleration at time $t = 4$ seconds.

Acceleration

$$\frac{ds}{dt} = 3t^2 + 4t - 1$$

$$\frac{d^2s}{dt^2} = 6t + 4$$

$$a = 6t + 4$$

Distance

$$\frac{ds}{dt} = 3t^2 + 4t - 1$$

$$s = \int (3t^2 + 4t - 1)dt$$

$$= t^3 + 2t^2 - t + c$$

when $t = 0$, $s = 0$

so $s = t^3 + 2t^2 - t$

Therefore at $t = 4$

$s = 4^3 + 2 \times 4^2 = 92$ units

$v = 3 \times 4^2 + 4 \times 4 - 1 = 63$ units/s

$a = 6 \times 4 + 4 = 28$ units/s^2

There are many other applications where integration is used. Any formulae required are given in the question.

EXERCISE 10A

1 A particle is moving in a straight line so that, at time t, its distance from a fixed point is x. Its velocity v is given by $v = t^2 + 4$ and, when $t = 0$, $x = 0$ units.
(i) Find expressions for the acceleration f and distance travelled x as functions of t.
(ii) Calculate how far the particle travelled during the third second of motion.

2 A particle is moving in a straight line so that, at time t, its distance from a fixed point is x. Its acceleration f is given by $f = k \cos^2 t$ and, when $t = 0$, $x = 0$ units and $v = 10$ units/s. Find expressions for the velocity v and distance travelled x as functions of t.

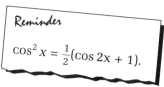

> **Reminder**
> $\cos^2 x = \tfrac{1}{2}(\cos 2x + 1).$

3 The diagram shows a sketch of the curve with equation $y = (x + 2)(x - 1)^2$.
Find the total shaded area.

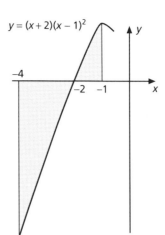

$y = (x + 2)(x - 1)^2$

4

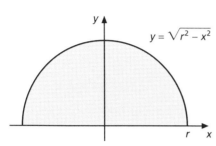

$y = \sqrt{r^2 - x^2}$

The diagram shows a semicircle with equation $y = \sqrt{r^2 - x^2}$. Using the substitution $x = a \sin t$, show that the area of the semicircle $= \frac{1}{2}\pi r^2$.

5 The semicircle shown in question **4** is rotated about the x-axis. Show that the volume of solid formed is $\frac{4}{3}\pi r^3$.

6 A car starts from rest and proceeds in a straight line. Its acceleration t seconds after its start is $\frac{1}{8}(30 - t)\,\text{m/s}^2$.

a Find the velocity after 4 seconds.

At this instant, acceleration ceases and the car continues to travel at constant velocity.

b Determine how far the car has travelled altogether after 16 seconds.

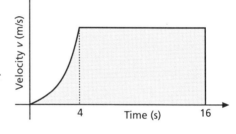

c The diagram shows the velocity–time graph for the car's movement. Calculate the total shaded area and compare this answer with your answer to **b**.

7 A particle, starting from rest, proceeds in a straight line. Its acceleration after t seconds is given by $a = 4 \sec^2 t$ units/s^2 where $0 \leq t \leq 1$. Calculate the velocity after 0.5 seconds and the distance travelled after $\frac{\pi}{4}$ seconds.

8 A particle proceeds in a straight line with acceleration given by $a = -e^{-4t}$. At A $t = 0$ and the velocity is 1. At B the velocity is zero. Calculate the distance between A and B.

9 Calculate the area enclosed by the curves $y = 3 \cos x$, $y = 1 + \cos x$ and the lines $x = 0$ and $x = \pi$.

10 a Find the area between the curve $y = e^{-x}$, the x-axis and the lines $x = 0$ and $x = z$.

b Write down the limiting value of the area as $z \to \infty$.

c In the same way, find the limiting value of the volume of revolution formed by rotating the curve $y = e^{-x}$, $x \geq 0$, about the x-axis.

11 The diagram shows an ellipse with equation

$\dfrac{x^2}{a^2} + \dfrac{y^2}{b^2} = 1$. If this curve is rotated about the x-axis,

a solid is formed called an ellipsoid.

Find an expression for the volume of this solid.

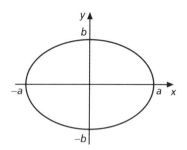

12

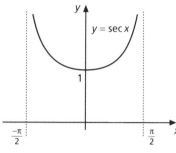

The diagram shows a sketch of the curve $y = \sec x$

in the range $-\dfrac{\pi}{2} < x < \dfrac{\pi}{2}$. Find the volume of

revolution obtained by rotating the section of the

curve between $x = -\dfrac{\pi}{4}$ and $x = \dfrac{\pi}{4}$ about the x-axis.

13 Using the substitution $3 - \sqrt{x} = t$, evaluate $\displaystyle\int_0^4 \dfrac{1}{3 - \sqrt{x}}\,\mathrm{d}x$.

14 Part of the curve $y = 4\sqrt{x}$ between $x = 0$ and $x = 2$ is rotated about the x-axis to form a paraboloid (like a car headlamp). Find the volume of revolution so obtained.

15 Show that $\displaystyle\int_0^{\frac{\pi}{4}} \sin^2 x \cos^3 x \,\mathrm{d}x = \dfrac{7\sqrt{2}}{120}$.

16 The diagram shows the straight line

$y = \dfrac{1}{4}x - \dfrac{1}{4}$ and the curve $y = \dfrac{1}{x} - \dfrac{1}{x^2}$.

Show that area enclosed between the

two curves is $\ln 2 - \dfrac{5}{8}$.

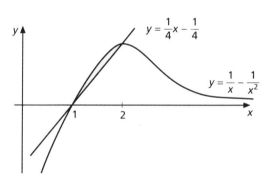

EXERCISE 10B

1 For an area bounded between a curve, the x-axis and, where necessary, lines $x = a$ and $x = b$, the coordinates of the centre of gravity are

$$\left(\dfrac{\displaystyle\int_a^b xy\,\mathrm{d}x}{\displaystyle\int_a^b y\,\mathrm{d}x}, \dfrac{\displaystyle\int_a^b y^2\,\mathrm{d}x}{2\displaystyle\int_a^b y\,\mathrm{d}x} \right).$$

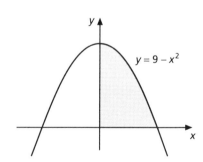

Show that the coordinates of the centre of gravity

of the area shown are $\left(\dfrac{9}{8}, \dfrac{18}{5} \right)$.

2 Using the formulae from question **1**, find the coordinates of the centre of gravity of

the area bounded by the curve $y = \dfrac{1}{x}$, the x-axis and the lines $x = 1$ and $x = 2$.

3 The diagram shows a sketch of the curve with equation $y = (x + 1)(x - 2)^2$.
Calculate the x-coordinate of the centre of gravity of the part shaded above the x-axis.
(Use the formula given in question **1**.)

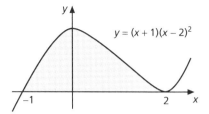

$y = (x + 1)(x - 2)^2$

4

The logo depicted is for a 'Healthier Teeth' campaign. The cartesian diagram shows the relevant mathematical information.
Calculate the area of the logo.

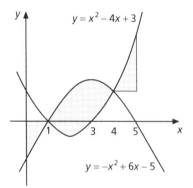

$y = x^2 - 4x + 3$

$y = -x^2 + 6x - 5$

5

$y = x - \dfrac{3x}{3 - x^2}$

$y = -\dfrac{1}{2}x$

The diagram shows the straight line $y = -\dfrac{1}{2}x$ and the curve $y = x - \dfrac{3x}{3 - x^2}$.
Show that the area enclosed between the two curves is $\dfrac{3}{2} + \ln\dfrac{8}{27}$.

6 Diagram 1 below shows the curve $x = \sqrt{r^2 - y^2}$. If this is rotated around the y-axis the solid of revolution formed is a hemisphere (diagram 2). However, if we rotate the part of the curve marked CD about the y-axis we obtain what is called the cap of a sphere (diagrams 3 and 4).
Let the cap have thickness d, as shown. Find a formula for the volume of this cap.

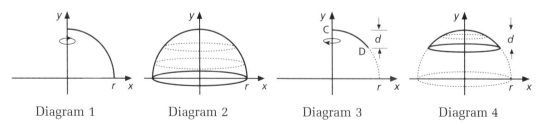

Diagram 1   Diagram 2   Diagram 3   Diagram 4

7 a Using the substitution $x = \dfrac{a + b}{2} + \dfrac{b - a}{2}\cos t$, evaluate $\displaystyle\int_a^b \sqrt{(x - a)(b - x)}\,dx$.

b Interpret your answer by examining a sketch of the function and the appropriate area.

8

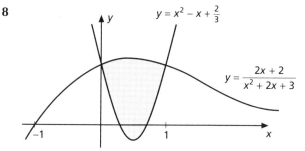

The diagram shows the curves $y = \dfrac{2x + 2}{x^2 + 2x + 3}$ and $y = x^2 - x + \dfrac{2}{3}$.

The curves intersect at $x = 0$ and $x = 1$.

Show that the area enclosed between the two curves is $\ln 2 - \dfrac{1}{2}$.

9 $y = \dfrac{x}{1 + x^2}$

  a Show that $\left(\dfrac{x}{y}\right)^2 - 2x^2 = 1 + x^4$ and that

$$\int_0^1 \left(\dfrac{1 - x^2}{(1 + x^2)\sqrt{1 + x^4}}\right) dx = \int_0^{\frac{1}{2}} \left(\dfrac{1}{\sqrt{1 - 2y^2}}\right) dy.$$

  b Using the substitution $y = \dfrac{1}{\sqrt{2}} \sin t$, evaluate $\displaystyle\int_0^{\frac{1}{2}} \left(\dfrac{1}{\sqrt{1 - 2y^2}}\right) dy.$

10

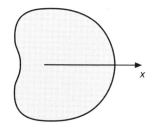

The diagram shows a limaçon. The polar equation of this curve is $r = 1 + \dfrac{1}{2}\cos t$ and the area enclosed by this curve is given $A = \dfrac{1}{2}\displaystyle\int_0^{2\pi} r^2 \, dt$. Calculate this area.

CHAPTER 3 REVIEW

1 Find the functions obtained from the following indefinite integrals.

a $\displaystyle\int\left(\frac{2x - x^3}{x^{\frac{3}{2}}}\right)dx$ **b** $\displaystyle\int(\cos(4x - 1))\,dx$ **c** $\displaystyle\int(2x + 3)^5\,dx$ **d** $\displaystyle\int(3x^2 - 1)^2\,dx$

e $\displaystyle\int(\sec^2 3x)\,dx$ **f** $\displaystyle\int(e^{4x - 3})\,dx$ **g** $\displaystyle\int\left(\frac{4}{x}\right)dx$ **h** $\displaystyle\int\frac{1}{(1 - 2\sin^2 x)^2}\,dx$

2 Find the following indefinite integrals using a substitution.

a $\displaystyle\int\left(x^2 + 2\right)\left(\frac{1}{3}x^3 + 2x\right)dx$ **b** $\displaystyle\int(-2\sin 2x e^{\cos 2x})\,dx$

3 Find $\displaystyle\int(x + 3)\sqrt[3]{x^2 + 6x - 1}\,dx$; let $x^2 + 6x - 1 = t$.

4 Find $\displaystyle\int\frac{1}{16 + x^2}\,dx$; let $x = 4\tan t$.

5 Show that $\displaystyle\int_0^1 4x\sqrt[3]{1 - x^2}\,dx = \left(\frac{3}{2}\right)$. Use the substitution $1 - x^2 = t^2$.

6 Show that $\displaystyle\int_1^{\sqrt{3}}\frac{1}{\sqrt{4 - x^2}}\,dx = \frac{\pi}{6}$ using the substitution $x = 2\sin t$.

7 Find $\displaystyle\int_a^{2a}\frac{4}{x}\,dx$.

8 By using appropriate trigonometric formulae, show that

$\displaystyle\int_{\frac{\pi}{6}}^{\frac{\pi}{3}}\frac{1}{\csc^2 x - 1}\,dx = \frac{3 - 1}{\sqrt{3}} - \frac{\pi}{6}$.

9 Calculate the shaded area shown in the diagram.

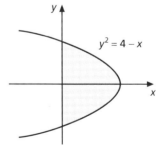

$y^2 = 4 - x$

10 A particle starts from rest and its acceleration a after t seconds is given by $a = 6t + 2$. Determine the velocity and the distance from the start after 3 seconds.

11 The section of the curve $y = x^3$ between $y = 0$ and $y = 8$ is rotated about the y-axis to form a volume of revolution. Calculate this volume V, given that $V = \displaystyle\int_a^b \pi x^2\,dy$.

CHAPTER 3 SUMMARY

1 If $F'(x) = f(x)$ then $F(x)$ is an antiderivative of $f(x)$.

2 If $F(x)$ is an antiderivative of $f(x)$ then

(i) $\int f(x)\,dx = F(x) + c$, where c is a constant of integration

(ii) $\int (af(x) + bf(x))\,dx = a\int f(x)\,dx + b\int g(x)\,dx$

(iii) $\int_a^b f(x)\,dx = F(b) - F(a)$

(iv) $\int_a^c f(x)\,dx = \int_a^b f(x)\,dx + \int_b^c f(x)\,dx,\ a < b < c$

(v) $\int_a^b f(x)\,dx = -\int_b^a f(x)\,dx$

3 Standard integrals include

(i) $\int x^n\,dx = \dfrac{1}{n+1}x^{n+1} + c$, where $n \neq -1$

(ii) $\int \cos x\,dx = \sin x + c$

(iii) $\int \sin x\,dx = -\cos x + c$

(iv) $\int e^x\,dx = e^x + c$

(v) $\int \dfrac{1}{x}\,dx = \ln|x| + c$, where $|x| = \begin{cases} x & \text{when } x \geq 0 \\ -x & \text{when } x < 0 \end{cases}$

(vi) $\int \sec^2 x\,dx = \tan x + c$

4 Special forms. If $F(x)$ is an antiderivative of $f(x)$ then

(i) $\int f(ax + b)\,dx = \dfrac{1}{a}F(ax + b) + c$

(ii) $\int \dfrac{f'(x)}{f(x)}\,dx = \ln|f(x)| + c$

(iii) $\int f'(x)f(x)\,dx = \dfrac{1}{2}[f(x)]^2 + c$

5 If $y = f(x)$ then $dy = f'(x)\,dx$ where dx and dy are the x-differential and y-differential respectively.

6 In the integral $\int g(f(x)).f'(x)\,dx$, let $u = f(x)$ then $du = f'(x)\,dx$ and the integral becomes $\int g(u)\,du$.

7 The area between the curve, the x-axis and the lines $x = a$ and $x = b$ is:

$$A = \int_{x=a}^{b} f(x)\,dx$$

Where the area straddles the x-axis, each portion must be computed separately.

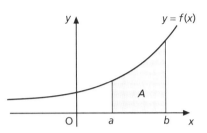

8

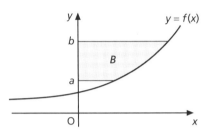

The area between the curve, the y-axis and the lines $y = a$ and $y = b$ is:

$$B = \int_{y=a}^{b} f^{-1}(y)\,dy$$

Where the area straddles the y-axis, each portion must be computed separately.

9 The volume of the solid generated when the curve $y = f(x)$, from $x = a$ to $x = b$, is revolved around the x-axis is:

$$V = \pi \int_{x=a}^{b} [f(x)]^2\,dx$$

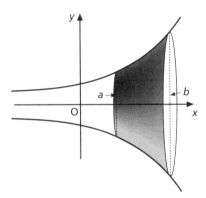

10

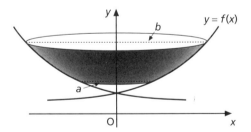

The volume of the solid generated when the curve $y = f(x)$, from $y = a$ to $y = b$, is revolved around the y-axis is:

$$V = \pi \int_{y=a}^{b} [f^{-1}(y)]^2\,dy$$

11 If $s(t)$ is the displacement of a particle from the origin, $v(t)$ its velocity and $a(t)$ its acceleration at time t, then:

(i) $v(t) = s'(t)$

(ii) $a(t) = v'(t) = s''(t)$

(iii) $s(t) = \displaystyle\int v(t)\,dt$

(iv) $v(t) = \displaystyle\int a(t)\,dt$

4 Properties of Functions

Historical note

Descartes

In 1637, René Descartes published a book which became known as *Discourse on Method*. Attached to the book are three appendices, one of which, *La géométrié*, explains his new invention: coordinate geometry. The main aim of this appendix was to bridge the gap between the two branches of mathematics: algebra and geometry. In *La géométrié* Descartes developed, for the first time, the concept of *the equation of a curve*.

Definitions

Reminders

A function, *f*, is defined as a rule which assigns each member of a set A uniquely to a member of a set B.

A function *f* assigns exactly one value *y* to each *x*. We write $y = f(x)$ or $f: x \rightarrow y$ ('*f* maps *x* to *y*').

y is referred to as the *image* of *x* in *f*.

Set A is referred to as the *domain* of the function and the set B as the *co-domain*.

The subset of B which is the set of all images of the function is called the *range* of the function.

It is possible for $f(a) = f(b)$ and yet $a \neq b$.

In this chapter certain sets are used often and are assigned specific symbols:

N, the set of natural numbers {1, 2, 3, 4, ...}
W, the set of whole numbers { 0, 1, 2, 3, 4, ...}
Z, the set of integers {...−3, −2, −1, 0, 1, 2, 3, ...}
Q, the set of rational numbers
R, the set of real numbers

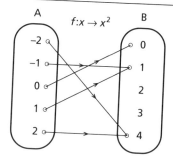

This example shows the function $f:x \rightarrow x^2$ linking each member of the set {−2, −1, 0, 1, 2} to members of the set {0, 1, 2, 3, 4}.

Note that, to meet with the definition of a function, each element of A has an image, and only one image.

The domain is {−2, −1, 0, 1, 2}.
The co-domain is {0, 1, 2, 3, 4}.
The range is {0, 1, 4}.
$f(-2) = f(2) = 4$ but $-2 \neq 2$

Often restrictions have to be placed on these sets to provide a suitable domain and range.

Example 1　Consider the function $f(x) = \sqrt{x}$.
What would be a suitable domain and range?

R would be an unsuitable domain since negative values of x have no image in R, but the set of positive real numbers and zero do form a suitable domain: $R^+ + \{0\}$.
R would be an unsuitable range, since it does not correspond to any suitable domain. Again $R^+ + \{0\}$ forms a suitable range for the given domain.

Example 2　Consider the function $f(x) = \dfrac{1}{(x-1)}$.
What would be a suitable domain and range?

Here the set of real numbers, except 1, provides a suitable domain: $R - \{1\}$.
The function can take all real values except 0, so the corresponding range is $R - \{0\}$.

EXERCISE 1

1 **(i)** Write down the largest suitable domain for each of the following functions.
　(ii) State the corresponding range where possible.

　a $f(x) = \sin x$　　**b** $f(x) = \sqrt{(x-2)}$　　**c** $f(x) = x!$ (x factorial)　　**d** $f(x) = \tan x$

　e $f(x) = x^2$　　　**f** $f(x) = 1 + \sin x$　　**g** $f(x) = \sqrt{(x^2)}$　　　**h** $f(x) = \dfrac{1}{\sin x}$

2

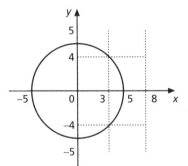

The diagram illustrates the relation $x^2 + y^2 = 5^2$.

　a What value(s) of y correspond to an x-value of: **(i)** 4　**(ii)** 8
　b Describe why $x^2 + y^2 = 5^2$ does not define a function on R.
　c Describe suitable restrictions which would allow you to work with this relationship as a function.

3 When a relationship is graphed, it is relatively easy to identify whether or not the relationship is a function on R. A relation is not a function if there exists a vertical line which:
　• cuts the graph more than once, or
　• does not cut the graph at all.
　Suitable restrictions are also then easily identified. ·

Which of these graphs could represent a function?

a
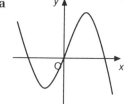

b
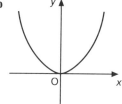

c
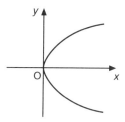

d
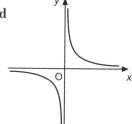

e
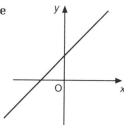

f
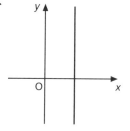

The modulus function, $y = |x|$

The modulus, or *absolute value,* of x is denoted by $|x|$ and is defined as

$$|x| = \begin{cases} x & \text{when } x \geq 0 \\ -x & \text{when } x < 0 \end{cases}$$

Examples:
$|2| = 2$; $|-3| = 3$; $|2.97| = 2.97$; $|-5.87| = 5.87$; $|\sin 30°| = 0.5$; $|\sin 210°| = 0.5$

The properties of $|x|$ include:

		Examples
1	$|x| = \sqrt{(x^2)}$	1 $|-2| = \sqrt{((-2)^2)} = \sqrt{4} = 2$
2	$|x + y| \leq |x| + |y|$	2 $|-3 + 2| \leq |-3| + |2| \ldots 1 \leq 3 + 2$
3	$|x \times y| = |x| \times |y|$	3 $|-3 \times 2| = |-3| \times |2|$
4	$|x| \leq a \Leftrightarrow -a \leq x \leq a$	4 $|5| \leq 8 \Leftrightarrow -8 \leq 5 \leq 8$
5	$|x| \geq a \Leftrightarrow -a \geq x \text{ or } x \geq a$	5 $|-2| \geq 1 \Leftrightarrow -1 \geq -2 \text{ or } -2 \geq 1$
		$\ldots$ in this case $-1 \geq -2$

The function $y = x$

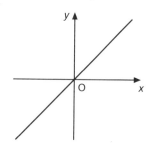

The function $y = |x|$

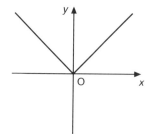

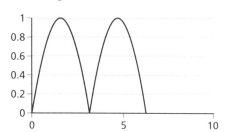

	A	B
1	x	abs(sin(x)
2	0	=ABS(SIN(A2))
3	=A2+PI()*0.1	=ABS(SIN(A3))
4	=A3+PI()*0.1	=ABS(SIN(A4))

Fill down to A30

Here a spreadsheet has been asked to graph $y = |\sin x|$.

Note that to draw $y = |f(x)|$ we need only reflect the negative portions of the graph $y = f(x)$ in the x-axis.

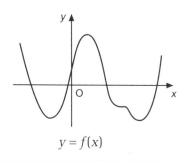

$$y = f(x)$$

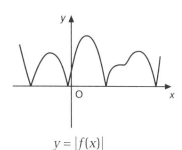

$$y = |f(x)|$$

EXERCISE 2

For each of these functions sketch the graph of $f(x)$ and $|f(x)|$.

1 $f(x) = x - 1$ 2 $f(x) = 2x - 1$ 3 $f(x) = \cos x°, 0 \leq x \leq 360$

4 $f(x) = x^2 - 1$ 5 $f(x) = 1 - x^2$ 6 $f(x) = x^2$

7 $f(x) = x^3$ 8 $f(x) = \ln x, x > 0$ 9 $f(x) = \tan x°, 0 \leq x \leq 360$

Inverse functions

A *one-to-one correspondence* is a special function where $a = b \Leftrightarrow f(a) = f(b)$.

In such a case the rule, f, which links each x in A to its image, $f(x)$, in B is reversible. A rule then exists which links the image, $f(x)$, in B back to the corresponding x in A. This rule is called *the inverse of the function*. It is denoted by f^{-1} and has the property that

$$y = f(x) \Leftrightarrow f^{-1}(y) = f^{-1}(f(x)) = x$$

For example, $f(x) = 2x + 1$ has an inverse $f^{-1}(x) = \dfrac{(x - 1)}{2}$.

Note that $f^{-1}(f(x)) = f^{-1}(2x + 1) = \dfrac{(2x + 1) - 1}{2} = \dfrac{2x}{2} = x$.

If the point (x, y) lies on the graph $y = f(x)$ then (y, x) lies on the graph $y = f^{-1}(x)$

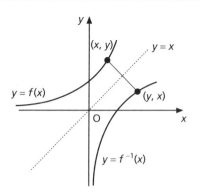

> *The graph of the inverse function is the reflection of the graph of the function in the line $y = x$*

If the inverse exists then it can be found by interchanging x and y in the formula $y = f(x)$, i.e. $x = f(y)$, and then making y the subject of the formula: $y = f^{-1}(x)$.

Placing restrictions on the domain and range can often ensure that a function has an inverse.

Example $f(x) = 4x^2 + 1$. Find $f^{-1}(x)$ and state a suitable domain and range for the function.

If the function can be represented by $y = 4x^2 + 1$, then its inverse, if it exists, can be represented by

$$x = 4y^2 + 1 \Rightarrow y = \frac{\sqrt{(x-1)}}{2} \qquad \text{taking the positive root}$$

$$\text{So } f^{-1}(x) = \frac{\sqrt{(x-1)}}{2}$$

Note $(x - 1) \geq 0$ if $f^{-1}(x)$ is real.

The largest suitable domain for this function is the set of numbers, x, such that $x \geq 1$ and x is real. The corresponding range is, therefore, the set of numbers, y, such that $y \geq 0$ and y is real. These sets of numbers can be written in a more compact fashion:

domain = $\{x : x \geq 1, x \in R\}$ and range = $\{y : y \geq 0, y \in R\}$
(The symbol '$\in$' stands for 'is a member of'.)

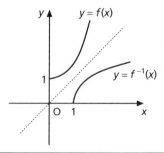

The domain and range of the inverse function give us the range and domain respectively of the original function:

domain = $\{x : x \geq 0, x \in R\}$ and range = $\{y : y \geq 1, y \in R\}$

The diagram illustrates the situation.

EXERCISE 3

1 Each of the following functions, $f(x)$, has an inverse.
   Find a formula for the inverse function $f^{-1}(x)$.

   a $f(x) = 3x + 4$
   b $f(x) = 5x - 1$
   c $f(x) = 3 - x$

   d $f(x) = 4 - 2x$
   e $f(x) = 8x^3$
   f $f(x) = 1 - x^5$

   g $f(x) = \dfrac{1}{x + 1}$
   h $f(x) = \dfrac{x}{x - 1}$
   i $f(x) = (x - 1)^{\frac{1}{3}}$

2 With suitable restrictions to the domain and range, each of these functions has an inverse. Find the inverse and state the largest suitable domain and range for the function.

a $f(x) = x^2$ **b** $f(x) = x^2 - 4$ **c** $f(x) = (x + 1)^2$

d $f(x) = (2x - 1)^2 - 1$ **e** $f(x) = x^2 + 6x - 1$ [Hint: complete the square.]

f $f(x) = \dfrac{1}{1 + x^2}$ **g** $f(x) = \sqrt{2x - 1}$

3 Each of the following is the graph of a function. Sketch the graph of its inverse.

a
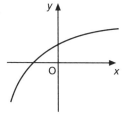

b
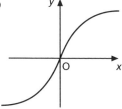

c
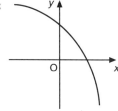

d
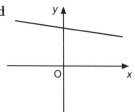

e
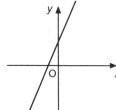

f
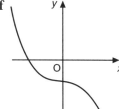

The sketches of question **3** illustrate the fact that for a function to have an inverse it must be either increasing over its domain or decreasing over its domain.

If we wish to find restrictions so that a function has an inverse then we must look for a suitable domain where this is happening.

Example

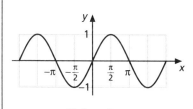

$f(x) = \sin x$

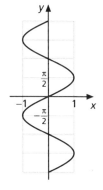

Reflect the curve in $y = x$.

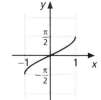

Find the largest region over which the curve is always increasing (or always decreasing).

$f(x) = \sin^{-1} x$

Domain $= \{x: -1 \leq x \leq 1, x \in R\}$

Range $= \left\{y: -\dfrac{\pi}{2} \leq y \leq \dfrac{\pi}{2}, y \in R\right\}$

EXERCISE 4

1 **a** Make a sketch of $y = \cos x$.

  b The inverse function $y = \cos^{-1} x$ has a domain of $-1 \le x \le 1$ and a range of $0 \le y \le \pi$. Make a sketch of $y = \cos^{-1} x$ to illustrate this.

2 **a** Sketch $y = \tan x$.

  b The inverse function $y = \tan^{-1} x$ has a domain of R and a range of $-\dfrac{\pi}{2} < y < \dfrac{\pi}{2}$. Make a sketch of $y = \tan^{-1} x$.

3 With the aid of calculators, spreadsheets or otherwise, investigate the domain and range of:

	A	B
1	=-PI()/2	
2	=A1+PI()*0.1	=1/COS(A2)

  a $y = \sec x$ **b** $y = \sec^{-1} x$

  c $y = \operatorname{cosec} x$ **d** $y = \operatorname{cosec}^{-1} x$

  e $y = \cot x$ **f** $y = \cot^{-1} x$

4 Each of the following functions is defined so that an inverse exists. For each:
  (i) make a sketch
  (ii) use it to help you sketch the inverse function
  (iii) state a suitable domain and range for the given function.

  a $f(x) = \sin 3x$ **b** $f(x) = 3 \sin x$ **c** $f(x) = 3 + \sin x$

  d $f(x) = \cos 2x$ **e** $f(x) = 2 \cos x$ **f** $f(x) = 2 + \cos x$

  g $f(x) = \sin x \cos x$ **h** $f(x) = \sin^2 x$ **i** $f(x) = \sin^2 x + \cos^2 x$

5 **a** Make a sketch of the function $f(x) = e^x$ marking on it a point to the left and to the right of the y-axis.

  b State the domain and range of the function.

  c **(i)** Sketch the inverse function.
      (ii) State its equation.
      (iii) Give the domain and range of the inverse.

6 For each of the following functions:
  (i) make a sketch
  (ii) use it to help you sketch the inverse function
  (iii) state a suitable domain and range for the given function.

  a $f(x) = e^{2x}$ **b** $f(x) = e^{-x}$ **c** $f(x) = e^x - 1$

  d $f(x) = \ln 2x$ **e** $f(x) = \ln (x^2)$ **f** $f(x) = 3 + \ln x$

7 With the aid of calculator, spreadsheet or otherwise, explore the domain and range of:

  a $f(x) = e^x + e^{-x}$ **b** $f(x) = e^x - e^{-x}$ **c** $f(x) = \dfrac{10 \ln x}{x}$

	A	B
1	-4	=EXP(A1)+EXP(-A1)
2	=A1+1	=EXP(A2)+EXP(-A2)
3	=A2+1	=EXP(A3)+EXP(-A3)

	A	B
1	1	=10*LN(A1)/A1
2	=A1+1	=10*LN(A2)/A2
3	=A2+1	=10*LN(A3)/A3

Sketching polynomials

Reminder

When sketching the graph of a polynomial one should try to ascertain certain facts:

1 Where does the curve cut the y-axis? ... Where $x = 0$
2 Where does the curve cut the x-axis? ... Where $f(x) = 0$
3 Where are the stationary points? ... Where $f'(x) = 0$
4. How does $f(x)$ behave as x approaches $\pm \infty$?

Example   Given $f(x) = x^3 - x^2 - x + 1$, sketch the graph of $y = f(x)$.

1 $x = 0 \Rightarrow y = f(0) = 1$ so the curve cuts the y-axis at $(0, 1)$

2 $y = 0 \Rightarrow x^3 - x^2 - x + 1 = 0$
   $\Rightarrow (x - 1)(x - 1)(x + 1) = 0$ factorising
   $\Rightarrow x = 1$ (twice) or $x = -1$ so the curve cuts the x-axis at $(1, 0)$ and $(-1, 0)$

3 $f'(x) = 3x^2 - 2x - 1$
   At stationary points, $f'(x) = 0$
   $\Rightarrow 3x^2 - 2x - 1 = 0$
   $\Rightarrow (3x + 1)(x - 1) = 0$
   $\Rightarrow x = -\frac{1}{3}$ or $x = 1$
   $f\left(-\frac{1}{3}\right) = \frac{32}{27}; f(1) = 0$

x	$\rightarrow$	$-\frac{1}{3}$	$\rightarrow$	1	$\rightarrow$
$3x + 1$	$-$	0	$+$	$+$	$+$
$x - 1$	$-$	$-$	$-$	0	$+$
$f'(x)$	$+$	0	$-$	0	$+$
Slope	╱	‾‾	╲	__	╱

A table of signs gives the nature of the stationary points: a maximum turning point at $\left(-\frac{1}{3}, \frac{32}{27}\right)$; a minimum turning point at $(1, 0)$.

4 For very large positive values of x, $f(x)$ will be very large and positive. The right-hand *tail* of the curve is in the first quadrant.
   For very large negative values of x, $f(x)$ will be very large and negative. The left-hand *tail* of the curve is in the third quadrant.
   [You need only consider the x^3 term of $f(x)$ to see this.]

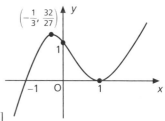

EXERCISE 5

1 Identify the stationary points of the following functions and determine their natures.
   a $2x^2 - 12x + 3$ b $x^2 - 6x - 5$ c $2 - 6x + 3x^2$
   d $2x^3 + 3x^2 - 12x + 1$ e $3 - 24x + 18x^2 - 4x^3$ f $x^4 - 8x^2 + 16$

2 Make sketches of the following polynomial functions.
Leave coordinates in surd form if necessary.

a $f(x) = x^4 - 4x^2$ **b** $f(x) = x^3 - 9x^2$ **c** $f(x) = x^3 - 12x - 16$

d $f(x) = (x - 9)(x - 1)(x - 6)$ **e** $f(x) = (x - 1)(x - 4)^2$

Extrema

Reminder

If a is in the domain of f:

1 points where $f'(a) = 0$ or $f'(a)$ does not exist are called *critical points*

2 a function has a local minimum at a if $f(a) \le f(x)$ for all x in some region centred at a

3 a function has a local maximum at a if $f(a) \ge f(x)$ for all x in some region centred at a

4 local extreme values (minima or maxima) occur at critical points (beware: not all critical points are local extrema – consider end-points)

5 a function has an end-point minimum at a if $f(a) \le f(x)$ for all x in a region close to a in the domain of f

6 a function has an end-point maximum at a if $f(a) \ge f(x)$ for all x in a region close to a in the domain of f

7 end-points are critical points

8 a function has a global minimum at a if $f(a) \le f(x)$ for all x in the domain of f

9 a function has a global maximum at a if $f(a) \ge f(x)$ for all x in the domain of f.

Example Consider the function $f(x) = |3 + 2x - x^2|$ defined in the domain $[0, 4)$, i.e. $0 \le x < 4$.

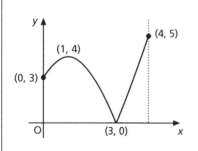

The critical points are $(0, 3)$, $(1, 4)$ and $(3, 0)$.

$(0, 3)$ is an end-point minimum since $f'(0)$ does not exist.

$(1, 4)$ is a local maximum point since $f'(1) = 0$.

$(3, 0)$ is a local minimum point since $f'(3)$ does not exist.

$(4, 5)$ is *not* an end-point maximum since 4 is not in the domain.

The global minimum is $(3, 0)$.
The function has no global maximum.

EXERCISE 6

1 Consider the nature of the end-points in the following functions.
    a $f(x) = x^2$ domain: $[-1, 2]$, i.e. $-1 \leq x \leq 2$
    b $f(x) = 4 - x^2$ domain: $[-1, 1]$
    c $f(x) = x^3 + x$ domain: $[-3, 0]$
    d $f(x) = 6 - x^3$ domain: $(0, 2]$, i.e. $0 < x \leq 2$
    e $f(x) = |2x|$ domain: $[-2, 3]$

2 Identify the critical points in the following functions.
    a $f(x) = x^2 + 2x + 1$ domain: $[-2, 3]$
    b $f(x) = \sqrt{(x + 1)}$ domain: $[-1, 3]$
    c $f(x) = \sqrt{(x^2 + 9)}$ domain: $[-4, 6)$
    d $f(x) = x^3 + x^2$ domain: $(-2, 2]$
    e $f(x) = x + \dfrac{1}{x}$ domain: $(0, 3]$

3 **(i)** Find the critical points.
   (ii) Identify end-points and local extrema.
   (iii) Hence identify the global maxima and minima where they exist.
    a $f(x) = x^2 - 4x - 5$ domain $[-2, 4]$
    b $f(x) = |\cos x|$ domain $\left[\dfrac{\pi}{4}, \dfrac{2\pi}{3}\right]$
    c $f(x) = |\ln x|$ domain $(0, e]$
    d $f(x) = 8x^3 - 3x^4$ domain $[-1, 3]$
    e $f(x) = \dfrac{x}{x^2 + 1}$ domain $[-2, 1)$

Concavity and points of inflexion

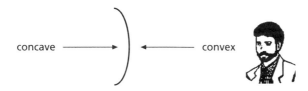

An observer looking at this curve from the left would say it was concave;
Looking at it from the right he would say it was convex

Examine this curve.

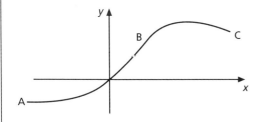

- In the region from A to B the gradient is increasing;
- between B and C the gradient is decreasing;
- at B the gradient stopped increasing and began decreasing.

In mathematics we call a region in which the gradient is increasing *concave up*, and one in which the gradient is decreasing *concave down*.

The derivative, $f'(x)$, measures the rate of change of the function $f(x)$ with respect to x (i.e. it measures the gradient).

The second derivative, $f''(x)$, measures the rate of change of the gradient with respect to x.

When this is positive, i.e. $f''(x) > 0$, the curve is concave up.

When this is negative, i.e. $f''(x) < 0$, the curve is concave down.

These facts can be used instead of a table of signs to determine the nature of stationary points.

> If $f'(a) = 0$
>
> then $f''(a) > 0$ implies there is a minimum turning point at $x = a$
>
> $f''(a) < 0$ implies there is a maximum turning point at $x = a$

When $f''(x) = 0$ then the derived function is stationary.

A point where the concavity changes from being in one state to the other is called a *point of inflexion*.

One should look for points of inflexion where $f''(x) = 0$ or where $f''(x)$ does not exist.

A table of signs of $f''(x)$ will confirm any change of concavity.

Example 1
$$f(x) = x^4$$
$$\Rightarrow f'(x) = 4x^3$$
$$\Rightarrow f''(x) = 12x^2$$
$$\Rightarrow f''(0) = 0$$

Is there a point of inflexion at $x = 0$?

x	$\rightarrow$	0	$\rightarrow$
$f''(x)$	+	0	+
Concavity	up		up

No change: no point of inflexion.

Example 2
$$f(x) = x^5$$
$$\Rightarrow f'(x) = 5x^4$$
$$\Rightarrow f''(x) = 20x^3$$
$$\Rightarrow f''(0) = 0$$

Is there a point of inflexion at $x = 0$?

x	$\rightarrow$	0	$\rightarrow$
$f''(x)$	–	0	+
Concavity	down		up

Change: there is a point of inflexion.

EXERCISE 7

1 Use the second derivative to prove the graph of

   a $f(x) = x^2$ is always concave up

   b $f(x) = \ln x$ is always concave down.

2 Discuss the concavity of the graph of

   a $f(x) = \sqrt{x},\ x \geq 0$

   b $f(x) = x + \dfrac{1}{x}$ in the interval $0 < x \leq 5$

   c $f(x) = \tan x$ in the interval $-\dfrac{\pi}{2} < x < \dfrac{\pi}{2}$

d $f(x) = x^3 + 6$

e **(i)** $f(x) = \sec x$ in the interval $-\dfrac{\pi}{2} < x < \dfrac{\pi}{2}$

    (ii) $f(x) = \sec x$ in the interval $\dfrac{\pi}{2} < x < \dfrac{3\pi}{2}$

3 **a** Identify the points of inflexion in the graph of $f(x) = \sin x$ in the interval $0 \le x \le 2\pi$.
    b In each case say whether the gradient of the tangent at that point is positive or negative.
    c Make a sketch of the curve over this interval and show the tangents.

4 Explore the concavity of $f(x) = \dfrac{1}{|x|}$ when   **a** $x < 0$ **b** $x > 0$.

5 Identify the points of inflexion of the following curves. In each case say whether the gradient of the tangent at that point is positive, negative or zero.
    a $f(x) = x^3$ **b** $f(x) = x^4 - 6x^2$
    c $f(x) = x^5 - 30x^3$ **d** $f(x) = \cot x$ in the interval $0 < x < \pi$

6 The zeros of a cubic function are x_1, x_2 and x_3. If $(a, f(a))$ is the point of inflexion of the cubic function, express a in terms of x_1, x_2 and x_3.

Odd and even functions

When sketching curves it can be useful to know of the existence of symmetries.

When $f(-x) = f(x)$ the curve is symmetrical about the y-axis. Such a function is known as an *even function*.

Example   Prove $f(x) = x^4 - 8x^2 + 3$ is an even function.

$$f(x) = x^4 - 8x^2 + 3$$
$$\Rightarrow f(-x) = (-x)^4 - 8(-x)^2 + 3$$
$$\Rightarrow f(-x) = x^4 - 8x^2 + 3$$
$$\Rightarrow f(-x) = f(x)$$

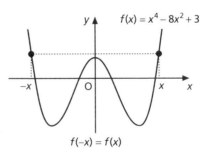

When $f(-x) = -f(x)$, the curve exhibits half-turn symmetry about the origin. Such functions are termed *odd functions*.

Example   Prove $f(x) = x^3 - 2x$ is an odd function.

$$f(x) = x^3 - 2x$$
$$\Rightarrow f(-x) = (-x)^3 - 2(-x)$$
$$\Rightarrow f(-x) = -x^3 + 2x$$
$$\Rightarrow f(-x) = -(x^3 - 2x)$$
$$\Rightarrow f(-x) = -f(x)$$

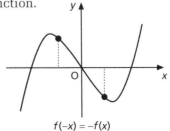

EXERCISE 8

1 a Prove that $2x^2 + 5$ is an even function.
  b Prove that $3x^5 + 7x^3 - 4x$ is an odd function.

2 For each of the functions below, say whether it is odd or even, and sketch enough of its graph to illustrate its symmetry.
  a $f(x) = x^6$ **b** $f(x) = x^5$ **c** $f(x) = \sin x$
  d $f(x) = \cos x$ **e** $f(x) = x^2 - 1$ **f** $f(x) = x + x^3$

3 Classify each of the following functions as odd, even or neither.
  a $f(x) = x$ **b** $f(x) = x^2 + x$ **c** $f(x) = \dfrac{x^2 + 1}{x^2}$

  d $f(x) = x + \dfrac{1}{x}$ **e** $f(x) = 1 - \dfrac{1}{x}$ **f** $f(x) = \sin x \cos x$

  g $f(x) = \sin x + \cos x$ **h** $f(x) = x^3 + x^2$ **i** $f(x) = e^{x^2}$

  j $f(x) = e^x + e^{-x}$ **k** $f(x) = e^x - e^{-x}$ **l** $f(x) = \ln x$

Vertical, horizontal and oblique asymptotes

$f(x)$ is a rational function if it can be expressed in the form $\dfrac{g(x)}{h(x)}$, where $g(x)$ and $h(x)$ are real polynomial functions and $h(x)$ is of degree 1 or greater.

If $h(a) = 0$, and the denominator of the rational function is $h(x)$, then the rational function is not defined at a. Then, as $x \to a$, $f(x) \to \pm\infty$ and the function is discontinuous at a.

A function is said to be continuous if $\lim\limits_{x \to a} f(x) = f(a)$

Example   How does $f(x) = \dfrac{1}{x - 1}$ behave in the neighbourhood of $x = 1$?

- As $x \to 1$, the denominator, $x - 1 \to 0$ and $f(x) \to \infty$.
- As $x \to 1$ from the *left*, $x - 1 < 0$ and the numerator, $1 > 0$, so $f(x) < 0$.
  To the left of the line $x = 1$, the curve $f(x) \to \infty^-$ (negative infinity).
- As $x \to 1$ from the *right*, $x - 1 > 0$ and the numerator, $1 > 0$, so $f(x) > 0$.
  To the right of the line $x = 1$, the curve $f(x) \to \infty^+$ (positive infinity).

The line $x = 1$ is called a *vertical asymptote* of the function.

The function is said to approach the line $x = 1$ asymptotically.

If $h(x)$ is the denominator of a rational function and $h(a) = 0$ then $x = a$ is a vertical asymptote of the rational function.

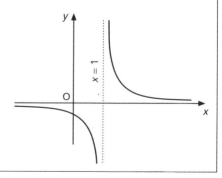

EXERCISE 9

1 For each function below,
  (i) identify the vertical asymptotes
  (ii) describe the behaviour of each function as it approaches its asymptotes from the left and from the right.

a $f(x) = \dfrac{x-1}{x+1}$

b $f(x) = \dfrac{1-x}{x+1}$

c $f(x) = \dfrac{3+x}{(x-1)(x+2)}$

d $f(x) = \dfrac{1}{x(2x+1)}$

e $f(x) = \dfrac{x-4}{(x-5)(x+1)}$

f $f(x) = \dfrac{x-1}{x^2-x-2}$

g $f(x) = \dfrac{x+3}{x^2-1}$

h $f(x) = \dfrac{1-x}{x^2-5x+6}$

i $f(x) = \dfrac{x+4}{x^3+1}$

2 Explore the behaviour of the following functions near their asymptotes.

a $f(x) = \tan x$

b $f(x) = \sec x$

c $f(x) = x^{-1}$

d $f(x) = \dfrac{1}{|x|}$

e $f(x) = \dfrac{1}{\ln(x)}$

f $f(x) = \dfrac{1}{e^x - 1}$

3 Investigate the behaviour of $f(x) = \dfrac{x}{\sqrt{(x^2)}}$.

Behaviour as $x \to \pm\infty$

If the function $f(x)$ can be expressed as $g(x) + \dfrac{m(x)}{n(x)}$, where $\dfrac{m(x)}{n(x)}$ is a proper rational function, then, as $x \to \infty$, $n(x) \to \infty$, $\dfrac{m(x)}{n(x)} \to 0$ and $f(x) \to g(x)$.

$f(x)$ is said to approach $g(x)$ *asymptotically*.

If $g(x)$ is a linear function, it is known as either a *horizontal asymptote* or an *oblique asymptote* depending on the gradient of the line $y = g(x)$.

The behaviour of the function as $x \to \infty^+$ and as $x \to \infty^-$ can be considered for each particular case.

Example 1   Find the asymptotes of $\dfrac{x+1}{x-1}$.

Vertical asymptote: denominator is $x - 1$
$$x - 1 = 0 \Rightarrow x = 1 \text{ is a vertical asymptote}$$

Other asymptote: $y = \dfrac{x+1}{x-1} = 1 + \dfrac{2}{x-1}$
          by polynomial division

As $x \to \infty$, $y \to 1 + 0$ so $y = 1$ is a *horizontal asymptote.*

As $x \to \infty^+$, $y \to 1 + 0^+$ so, to right, $f(x)$ approaches the asymptote from above.

As $x \to \infty^-$, $y \to 1 + 0^-$ so, to left, $f(x)$ approaches the asymptote from below.

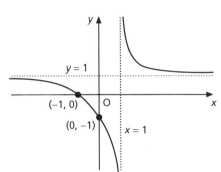

Example 2   Explore the asymptotes of $f(x) = \dfrac{2x^2 + x - 3}{x + 1}$.

Vertical asymptote: denominator is $x + 1$

$$x + 1 = 0 \Rightarrow x = -1 \text{ is a vertical asymptote}$$

Other asymptote: $f(x) = 2x - 1 - \dfrac{2}{x + 1}$ by polynomial division

As $x \to \infty$, $y \to 2x - 1 - 0$ so $y = 2x - 1$ is an *oblique asymptote*.
As $x \to \infty^+$, $y \to 2x - 1 - 0^+$ so, to right, $f(x)$ approaches the asymptote from below.
As $x \to \infty^-$, $y \to 2x - 1 - 0^-$ so, to left, $f(x)$ approaches the asymptote from above.

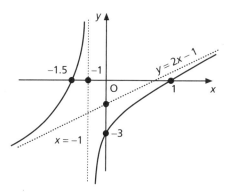

Note   The curve cuts *y*-axis when $x = 0$, $(0, -3)$; cuts *x*-axis when $y = 0$, $(1, 0)$
and $(-1.5, 0)$; and has stationary points when $f'(x) = 0$: no SPs

EXERCISE 10

1 For each function below:
   (i)   give the equation of any vertical asymptotes
   (ii) perform polynomial divisions as appropriate
   (iii) state the non-vertical asymptotes.
   (iv) consider the behaviour of the function as it approaches these asymptotes.

   a $f(x) = \dfrac{1}{x}$

   b $f(x) = \dfrac{x + 1}{x}$

   c $f(x) = \dfrac{x^2 + 2}{x}$

   d $f(x) = \dfrac{x - 1}{x + 1}$

   e $f(x) = \dfrac{1 - x}{x + 1}$

   f $f(x) = \dfrac{x^2 + 1}{x + 1}$

   g $f(x) = \dfrac{x^2 + 1}{x^2 - 1}$

   h $f(x) = \dfrac{x}{x^2 + 1}$

   i $f(x) = \dfrac{x - 1}{x^2 - x - 2}$

   j $f(x) = \dfrac{x^2 + 3}{x - 1}$

   k $f(x) = \dfrac{2x^2 + 2x - 3}{x + 2}$

   l $f(x) = \dfrac{x^3 - 3x}{x^2 + 1}$

2 Given that $f(x) = g(x) + \dfrac{m(x)}{n(x)}$ where $\dfrac{m(x)}{n(x)}$ is a proper rational function, we know the non-vertical asymptote is $y = g(x)$.

The curve $y = f(x)$ cuts the curve $y = g(x)$ where $\dfrac{m(x)}{n(x)} = 0$.

Find where each of the following cut their own asymptote.

a $f(x) = \dfrac{x^3 + 2x}{x^2 + 1}$

b $f(x) = \dfrac{2x^3 + x^2 + x + 1}{x^2}$

3 a Show that $\dfrac{x^3 - x^2 + 1}{x - 1}$ starts behaving like $y = x^2$ as $x \to \infty$.

b Comment on the behaviour of $\dfrac{x^3 - x^2 + x}{x - 1}$ as $x \to \infty$.

Bringing it all together

When sketching curves, gather as much information as possible from the list below:

1 Intercepts
   a Where does the curve cut the y-axis? $x = 0$
   b Where does the curve cut the x-axis? $y = 0$

2. Extras
   a Where is the function increasing? $f'(x) > 0$
   b Where is the function decreasing? $f'(x) < 0$

3 Stationary points
   a Where is the curve stationary? $f'(x) = 0$
   b Where is the function concave up? [for minimum] $f''(x) > 0$
   c Where is the function concave down? [for maximum] $f''(x) < 0$
   d Are there any horizontal points of inflexion? $f''(x) = 0$

4 Extras
   Where are the non-horizontal points of inflexion? $f''(x) = 0$

5 Asymptotes
   a Where are the vertical asymptotes?   denominator $= 0$
   b Where are the non-vertical asymptotes? $x \to \pm \infty$

A table of signs can be helpful to add detail.

EXERCISE 11

1 Sketch the following graphs

a $y = \dfrac{1}{x + 3}$

b $y = \dfrac{3}{2x + 8}$

c $y = \dfrac{x}{x + 2}$

d $y = \dfrac{x - 1}{x + 1}$

e $y = \dfrac{1 - x}{1 + x}$

f $y = \dfrac{x - 1}{x(x + 1)}$

g $y = \dfrac{x}{(x - 1)(x + 1)}$

h $y = \dfrac{x^2}{x + 1}$

i $y = x - \dfrac{1}{x}$

j $y = \dfrac{x^2}{1 - x}$

k $y = \dfrac{(2x + 3)(x - 6)}{(x + 1)(x - 2)}$

l $y = \dfrac{1}{(x - 2)(x - 4)}$

m $y = \dfrac{x^2 - x}{2x + 1}$

n $y = \dfrac{(x - 1)(x + 2)}{x - 2}$

o $y = \dfrac{x^2 - 3x - 10}{x - 2}$

2 Make use of the fact that the function is either odd or even to help you to draw its graph.

a $y = \dfrac{3}{x^2 - 3}$

b $y = \dfrac{x^2 - 9}{x}$

c $y = \dfrac{x^2 + 1}{x^2 - 1}$

3 a By examining the extrema of the function $f(x) = \dfrac{x}{x^2 + 1}$, prove that

$$-\dfrac{1}{2} \le \dfrac{x}{x^2 + 1} \le \dfrac{1}{2}$$

[Note that $x^2 + 1 > 0$ for all x.]

b **(i)** Is the function odd or even?

(ii) Sketch the graph $y = f(x)$.

4 a Prove that $2x^2 - x + 2 > 0$ for all x.

b Given that $f(x) = \dfrac{2x^2 + x + 2}{2x^2 - x + 2}$ and that $a \le f(x) \le b$ for all x, find the values of a and b.

c Sketch the graph of the function.

5 a Following similar logic to questions **3** and **4**, a student, studying the function $f(x) = \dfrac{3x^2 - 3}{6x - 10}$, came to the conclusion that $f(x) \ge 3$ and $f(x) \le \dfrac{1}{3}$. What had he ignored?

b Make a sketch of the function to illustrate the function.

6 a Show that the range of the function $f(x) = \dfrac{2x - 1}{2x^2 - 4x + 1}$ is R.

b Sketch the graph of the function.

Graphs of related functions

Reminder

Given the graph of $y = f(x)$

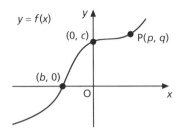

we can draw the graphs of
some related functions:

1. by reflection ...

| ... in the x-axis | ... in the y-axis | ... in the line $y = x$ | ... of negative f in the x-axis |

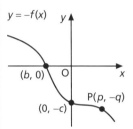
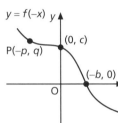
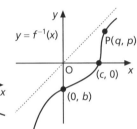
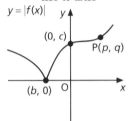

2. by translation ...

... in the x-direction   ... in the y-direction

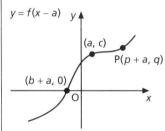
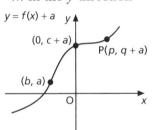

3. by scaling ...

        ... in the x-direction   ... in the y-direction

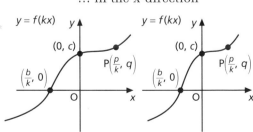
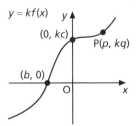
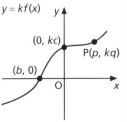

x-stretch if $0 < k < 1$ *x*-squash if $k > 1$ *y*-stretch if $k > 1$ *y*-squash if $0 < k < 1$

... or any combination of the above.

EXERCISE 12A

1 Given that all the graphs below are the result of simple transformations of the graph of $f(x) = x^2$, suggest a possible equation for each.
The images of the points (3, 9) and (0, 0) in the transformation are given.

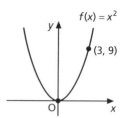

a 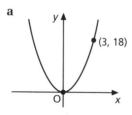 **b** 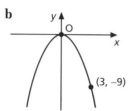 **c** 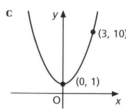 **d**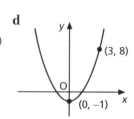

e 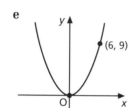 **f** 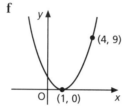 **g** 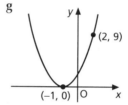 **h**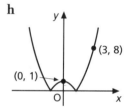

i 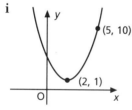 **j** 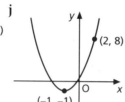 **k** 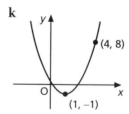 **l**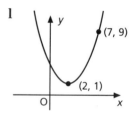

2 a Draw simple sketches of the graphs of the following functions.
State the coordinates of the vertex of the parabola in each case.
  (i) $f(x) = x^2 + 3$ **(ii)** $f(x) = (x - 2)^2$ **(iii)** $f(x) = (x + 3)^2 + 1$
  (iv) $f(x) = |x^2 - 3|$ **(v)** $f(x) = |(x - 2)^2|$ **(vi)** $f(x) = 2(x + 3)^2$
b Make a sketch of the inverse function of each, quoting a suitable range.

3 Knowing the graph $y = x^3$, sketch the graphs of the following, stating the coordinates of the point of inflexion in each case.
  a $-x^3$ **b** $x^3 + 2$ **c** $x^3 - 2$ **d** $(x + 1)^3$
  e $(-x)^3$ **f** $1 - x^3$ **g** $|x^3|$ **h** $x^{\frac{1}{3}}$
  i $3x^3$ **j** $(x - 3)^3 + 2$ **k** $|(x - 1)^3 + 1|$ **l** $(x + 1)^{\frac{1}{3}}$

4 Sketch each pair of graphs on the same diagram.
  a (i) $y = e^{-x}$ **(ii)** $y = e^{-x} + 2$ **b (i)** $y = e^x$ **(ii)** $y = e^{x-2}$
  c (i) $y = \ln x$ **(ii)** $y = 5 + \ln x$ **d (i)** $y = \ln x$ **(ii)** $y = \ln(x - 3)$
  e (i) $y = e^{-x}$ **(ii)** $y = \ln x^{-1}$ **f (i)** $y = e^{-x} + 2$ **(ii)** $y = \ln\left(\dfrac{1}{x - 2}\right)$

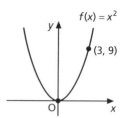

5 The diagram is a sketch of $y = f(x)$. Make sketches of the graphs of the following related functions marking the images of the given points on your diagram.

    a $y = f(x) + 2$ **b** $y = -f(x)$ **c** $y = -f(x) + 2$

    d $y = f(x + 2)$ **e** $y = 2f(x)$ **f** $y = 2f(x) + 2$

    g $y = f\left(\dfrac{x}{2}\right)$ **h** $y = 2f\left(\dfrac{x}{2}\right)$ **i** $y = 2f\left(\dfrac{x}{2}\right) + 2$

6 Make sketches of the following functions which are related to the function $y = g(x)$ whose graph is illustrated on the right. Give the coordinates of the points corresponding to those shown.

    a $y = g(x) + 2$ **b** $y = g(x + 2)$ **c** $y = -g(x)$

    d $y = g(-x)$ **e** $y = |g(x)|$ **f** $y = 3g(x)$

    g $y = g(2x)$ **h** $y = 3g(2x)$ **i** $y = 2g(x) + 2$

7 Sketch the following functions which are related to the function whose graph is as illustrated. Mark points as you feel appropriate.

    a **(i)** $y = h(x - 3)$ **(ii)** $y = h(x + 3)$

         (iii) $y = -h(x - 3)$ **(iv)** $y = h(-x) + 6$

         (v) $y = 2h(3x)$ **(vi)** $y = 2h(3x + 1)$

         (vii) $y = -2h(x - 1)$ **(viii)** $y = |h(x) - 1|$

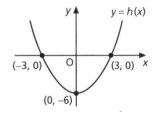

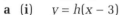

    b

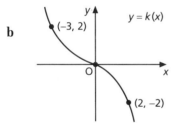

         (i) $y = -k(x)$ **(ii)** $y = -k(-x)$

         (iii) $y = 4 - k(-x + 1)$ **(iv)** $y = k^{-1}(x)$

         (v) $y = |k^{-1}(x)|$ **(vi)** $y = 3 - k^{-1}(x)$

    c **(i)** $y = -s(x)$ **(ii)** $y = 1 + s(-x)$

         (iii) $y = 2s(-x)$ **(iv)** $y = -1 - s(x + 2)$

         (v) $y = |s(x)|$ **(vi)** $y = 3 - s(3x)$

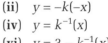

    d

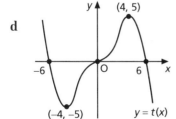

         (i) $y = -t(x)$ **(ii)** $y = -t(-x)$

         (iii) $y = 5 + t(6 - x)$ **(iv)** $y = t(x - 6)$

         (v) $y = |t(x - 6)|$ **(vi)** $y = 3 - 2t(4x)$

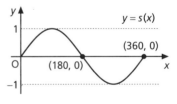

8 **a** Copy and complete the curve $y = u(x)$ given that $u(x)$ is an even function.

    b Sketch

         (i) $y = -u(x)$ **(ii)** $y = u(x) + 2$

         (iii) $y = 2 - u(x + 4)$ **(iv)** $y = |u(x)|$

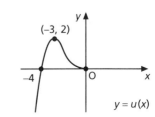

1 Part of the graph of the function $f(x) = x^{-1}$ is given.
   On separate diagrams make sketches of:

   a $y = (-x)^{-1}$

   b $y = \dfrac{x+2}{x} = 1 + \dfrac{2}{x}$

   c $y = \dfrac{2x-1}{x}$

   d $y = \dfrac{1}{|x|}$

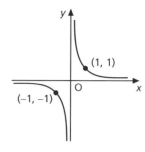

2 By expressing each in the form $a(bx - c)^2 + d$, and using the
   graph $y = x^2$ as a basis, sketch the following functions.

   a $f(x) = x^2 + 2x + 3$

   b $f(x) = x^2 + 6x - 1$

   c $f(x) = 3 + 2x - x^2$

   d $f(x) = |x^2 + 6x - 1|$

   e $f(x) = |3 + 2x - x^2|$

   f $f(x) = x^2 - 8x$

3 On the right is a sketch of the function $g(x) = x^3 - 3x^2 + 3x$.
   Based on this, make sketches of:

   a $h(x) = x^3 - 3x^2 + 3x + 1$

   b $m(x) = -x^3 - 3x^2 - 3x$ [Hint: $= g(-x)$]

   c $n(x) = 4 - x^3 - 3x^2 - 3x$

   d $p(x) = |2x^3 - 6x^2 + 6x|$

   e $p(x) = -x^3 + 3x^2 - 3x$

   f $g^{-1}(x)$ $[= (x-1)^{\frac{1}{3}} + 1]$

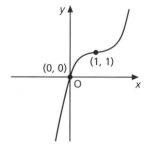

4 **a** Prove that the function $f(x) = x + \sin x$ is an increasing
         function.

   b A part of the graph of $f(x)$ is shown: from $x = -\pi$ to $x = \pi$.
         Use this to help you sketch the following:

         (i) $g(x) = 2x + 2\sin x \cos x$

         (ii) $h(x) = -x - \sin x$

         (iii) $k(x) = x + \pi + \sin(x + \pi)$

         (iv) $f^{-1}(x)$

5 Make sketches of the following functions.

   a $e^{(2x+1)}$

   b $-e^{(-x+1)}$

   c $|5 - e^{-x}|$

   d $\ln(x - 2)$

   e $3 - \ln(x - 1)$

   f $|\ln(x - 1)|$

6 *Investigation.* How are the graphs $y = f(x)$ and $y = \dfrac{1}{f(x)}$ related?
   Consider features such as

   • zeros of $f(x)$

   • stationary points and their natures

   • asymptotes.

   Graphics calculators can be useful here:
   define Y1 by typing in a suitable $f(x)$;
   define Y2 as 1/Y1.

CHAPTER 4 REVIEW

1 a State the domain and range of the function $f(x) = x!$

2 A function is defined in the domain $(0, 2]$ as $f(x) = 2x^2 - 6x + 3$.
  a Identify the critical points of the function.
  b Find the local maxima and minima where they exist.
  c What is the global maximum value of the function?

3 Describe the concavity of the function $f(x) = (x + 3)^3 + 1$
  and identify the point of inflexion.

4 a Make a sketch of the function $f(x) = \sin^{-1} x$.
  b State a suitable domain and range.

5 Draw sketches of the graphs of the following rational functions.

  a $\dfrac{x - 2}{x + 1}$
  b $\dfrac{x + 3}{x^2 + x - 2}$
  c $\dfrac{x^2}{x - 3}$

  d $\dfrac{(x - 3)(x + 6)}{(x - 1)(x + 2)}$
  e $\dfrac{x^2 + 2x - 3}{x + 1}$

6 Based on the graph of $y = f(x)$ shown on the right,
  make sketches of:

  a $y = 2f(x)$
  b $y = f(-x)$
  c $y = -f(x)$
  d $y = f(x) - 2$
  e $y = f(x + 4)$
  f $y = -f(3x)$
  g $y = 2f(3x + 2) - 1$
  h $y = |f(x)|$

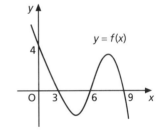

7 For the graph of $y = g(x)$ shown below, sketch the graphs:
  a $y = |g(x)|$
  b $y = h(x)$ where $h(x) = g^{-1}(x)$

  Comment on
  the graph of $h^{-1}(x)$.

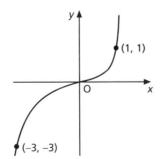

8 Sketch
  a $e^{ax + b}$
  b $\ln(ax + b)$ where a and b are constants.

CHAPTER 4 SUMMARY

1 A function, f, is a rule which links each member of a set A uniquely to a member of a set B.

2 Set A is referred to as the *domain* of the function and the set B as the *co-domain*. The subset of B which is the set of all images of the elements of the domain is called the *range* of the function.

3 If a is in the domain of f:
   (i)   points where $f'(a) = 0$ or $f'(a)$ does not exist are called critical points
   (ii) a function has a local minimum at a if $f(a) \leq f(x)$ for all x in some region centred at a
   (iii) a function has a local maximum at a if $f(a) \geq f(x)$ for all x in some region centred at a
   (iv) local extreme values (minima or maxima) occur at critical points
   (v)   a function has an end-point minimum at a if $f(a) \leq f(x)$ for all x in a region close to a in the domain of f
   (vi) a function has an end-point maximum at a if $f(a) \geq f(x)$ for all x in a region close to a in the domain of f
   (vii) end-points are critical points
   (viii) a function has a global minimum at a if $f(a) \leq f(x)$ for all x in the domain of f
   (ix) a function has a global maximum at a if $f(a) \geq f(x)$ for all x in the domain of f.

4 (i)   If $f''(a) > 0$ the curve is concave up at $x = a$.
   (ii) If $f''(a) < 0$ the curve is concave down at $x = a$.
   (iii) If $f'(a) = 0$ then: $f''(a) > 0$ implies there is a minimum turning point at $x = a$
                              $f''(a) < 0$ implies there is a maximum turning point at $x = a$.

5 Trigonometric functions and their inverses

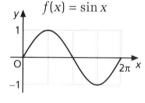

$f(x) = \sin x$

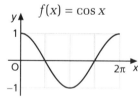

$f(x) = \cos x$

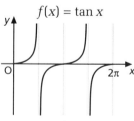

$f(x) = \tan x$

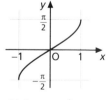

$f(x) = \sin^{-1} x$
domain: $[-1, 1]$
range: $\left[-\dfrac{\pi}{2}, \dfrac{\pi}{2} \right]$

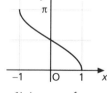

$f(x) = \cos^{-1} x$
domain: $[-1, 1]$
range: $[0, \pi]$

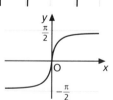

$f(x) = \tan^{-1} x$
domain: $(-\infty, \infty)$
range: $\left(-\dfrac{\pi}{2}, \dfrac{\pi}{2} \right)$

6 The exponential function and its inverse

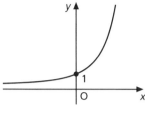

$$f(x) = e^x$$

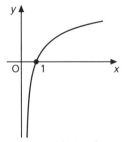

$$f(x) = \ln x$$

7 Symmetries
  (i) A function is said to be *even* if it is symmetrical about the *y*-axis: $f(x) = f(-x)$.
  (ii) A function is said to be *odd* if it is symmetrical about the *origin*: $f(x) = -f(-x)$.

8 Transformations
  From the graph $y = f(x)$, we get
  (i) $y = -f(x)$ by reflection in the *x*-axis
  (ii) $y = f(-x)$ by reflection in the *y*-axis
  (iii) $y = f^{-1}(x)$, the inverse function, by reflection in the line $y = x$
  (iv) $y = |f(x)|$, which equals $\sqrt{(f(x)^2)}$, by reflection of negative f in the *x*-axis
  (v) $y = f(x - a)$ by a translation of a units in the *x*-direction
  (vi) $y = f(x) + a$ by a translation of a units in the *y*-direction
  (vii) $y = f(kx)$ by scaling the *x*-coordinates by a factor of $\dfrac{1}{k}$
  (viii) $y = kf(x)$ by scaling the *y*-coordinates by a factor of k.

9 Asymptotes
  A function $f(x)$ which can be expressed in the form $g(x) + \dfrac{m(x)}{n(x)}$:
  (i) has a vertical asymptote $x = a$ where $n(a) = 0$
  (ii) has a non-vertical asymptote $y = g(x)$ where $\dfrac{m(x)}{n(x)}$ is a proper rational
    function and the orders of m and n differ by no more than 2
  (iii) will cut its non-vertical asymptote when $m(x) = 0$.

5 Systems of Equations

Systems of equations, redundancy and inconsistency

Reminders

Equations and solutions

An equation is a mathematical sentence, containing '=', whose truth depends on the value(s) of the variable(s) in the sentence.

The values of variables which make the sentence true are called *solutions* of the equation.

The equation $3x + 2y = 5$ has infinitely many solutions: for example $(1,1)$, $(9, -11)$, ... In fact, for any value of x, $\left(x, \dfrac{(5 - 3x)}{2}\right)$ is a solution.

This type of equation is called a *linear equation*: all solutions, when plotted on the cartesian plane, lie in a straight line.

Example   In 1973:
two second class and three first class stamps cost a total of 21 pence,
three second class and two first class stamps cost a total of 19 pence.
What was the cost of one of each?

Situations like this lead to systems of linear equations:

$$2x + 3y = 21 \qquad \text{equation 1}$$
$$3x + 2y = 19 \qquad \text{equation 2}$$

where x represents the cost of a second class stamp and y the cost of a first class stamp in pence.

Finding the solution is a fairly standard procedure:

Equation 1 × 3:	$6x + 9y = 63$	equation 3
Equation 2 × 2:	$6x + 4y = 38$	equation 4
Equation 4 − Equation 3:	$-5y = -25$	equation 5
Equation 5 ÷ (−5):	$y = 5$	

Substitute this value back into equation 3: $6x + 45 = 63$
$$\Rightarrow \quad 6x = 18$$
$$\Rightarrow \quad x = 3$$

The cost of a second class stamp was 3 pence;
The cost of a first class stamp was 5 pence.

This system of equations is known as a 2 × 2 system of linear equations because there are two linear equations and two variables.

Not every pair of 2 × 2 linear equations produces a unique solution like the above example, as the following two sets of equations show.

$3x + 3y = 6$
$4x + 4y = 8$

These equations are equivalent and, when any attempt to eliminate a variable is undertaken, the resulting unhelpful truism occurs: $0 = 0$.

One of the equations is said to be *redundant*.

There is in fact an infinite number of solutions, that is: for any value of x, $(x, 2 - x)$ is a solution.

$x + 4y = 6$
$2x + 8y = 10$

If we attempt to eliminate a variable by doubling equation 1, we get $0 = 2$, which is clearly impossible.

The equations are said to be *inconsistent*.

There are no solutions.

EXERCISE 1

1 For each of the following systems of equations either
   • solve it for its unique solution
   • declare one of the equations redundant and give the solution in the form $(x, ax + b)$ where a and b are constants, or
   • declare the system inconsistent and that there are no solutions.

   a $2x + y = 6$
       $x + 2y = 9$

   b $2x + 3y = 9$
       $10x + 6y = 16$

   c $4x + y = 3$
       $8x + 2y = 6$

   d $x - 3y = 2$
       $6y - 2x = 1$

   e $3x + 2y = 4$
       $2x + 5y = -1$

   f $6x = 2y - 4$
       $y = 2 - 3x$

In this system of 3 × 2 equations, the third equation is *redundant* since the solution, which does satisfy it, can be deduced from the first two equations only.
   $x + 2y = 7$
   $2x + y = 5$
   $7x + 8y = 31$
A redundant equation can be *built* from the sum of multiples of the other equations in the system.

In this system of 3 × 2 equations, the third equation is *inconsistent* with the other two since a solution based on the first two does not satisfy the third.
   $x + 2y = 7$
   $2x + y = 5$
   $7x + 8y = 30$
The system has no solution.

2 In each of the following systems, decide the relationship between the third equation and the other two. State the solution where appropriate.

a $2x + y = 7$
$x + 3y = 6$
$2x + 3y = 9$

b $x - 2y = -5$
$x + y = 1$
$3x + 4y = 4$

c $x + 3y = 2$
$3x + y = 3$
$x + 5y = 1$

d $4x - 2y = -2$
$x - 3y = 2$
$2x - y = -1$

3 Comment on this system of equations.
$$2x + 3y = 5$$
$$4x + 6y = 10$$
$$10x + 15y = 25$$

4 In the Greasy Spoon Café, an order of 1 cola and 3 teas cost £3.60. The bill for an order of 2 colas and 2 teas came to £4. A third order, for 3 colas and 2 teas, cost £5.

a Form a system of equations.

b Use the first two equations to find prices which satisfy both.

c Comment on the third order.

2 × 2 systems and matrices; the augmented matrix

In this chapter we will be studying systems of equations of different sizes, so we need a more compact way of organising and recording our working. To meet this need, we introduce the *matrix*.

A matrix (plural *matrices*) is simply a table of values organised in rows and columns. It is standard notation to contain the table within brackets. For example,

$$\begin{pmatrix} 2 & 4 \\ 1 & 5 \end{pmatrix}$$

is referred to as a 2 × 2 matrix. It has two rows and two columns.
Row 1 contains 2 and 4 as elements.
Column 2 contains 4 and 5 as elements.
The matrix contains the four elements 2, 4, 1 and 5.

It is common to assign an upper case letter in naming a matrix and to use the equivalent lower case letter to refer to the elements. Subscripts make the references unambiguous. For example,

$$A = \begin{pmatrix} a_{11} & a_{12} \\ a_{21} & a_{22} \end{pmatrix}$$

where a_{ij} refers to the element in the ith row and jth column.

The system of equations used in the postage stamp problem can be represented by matrices written in the form $A\mathbf{x} = \mathbf{b}$:

$$\begin{pmatrix} 2 & 4 \\ 1 & 5 \end{pmatrix} \begin{pmatrix} x \\ y \end{pmatrix} = \begin{pmatrix} 42 \\ 57 \end{pmatrix}$$

where A is a 2 × 2 matrix for the coefficients of the variables, column 1 for the x coefficients and column 2 for the y coefficients;

x is a 2×1 matrix for the variables, commonly called a vector in this context;
b is a 2×1 matrix for the constants on the right-hand side.

An even more compact representation is attained by combining matrix A and b into a bigger matrix known as the *augmented matrix*:

$$\begin{pmatrix} 2 & 4 & 42 \\ 1 & 5 & 57 \end{pmatrix}$$

This contains all the data needed to calculate a solution.

EXERCISE 2

1 Rewrite each system of equations:
   (i) in the matrix form $Ax = b$
   (ii) in the augmented matrix form.

   a $x + 2y = 4$
       $3x - y = 5$

   b $2x + 4y = 8$
       $x - 2y = -4$

   c $3x - 2y = 7$
       $x - y = 2$

   d $3x = -6$
       $4x + y = -7$

   e $0.5x + 2y = 3$
       $x - 1.5y = -5$

   f $0.4x + 0.6y = 0.2$
       $0.1x - 0.2y = 0.4$

2 Write out the system of equations in full which are represented by each of these augmented matrices.

   a $\begin{pmatrix} 2 & 1 & 3 \\ 3 & 4 & 7 \end{pmatrix}$
   b $\begin{pmatrix} 1 & 2 & 0 \\ 3 & 1 & 5 \end{pmatrix}$
   c $\begin{pmatrix} 1 & 1 & 7 \\ 1 & 0 & 4 \end{pmatrix}$
   d $\begin{pmatrix} 1 & 0 & 5 \\ 0 & 1 & 2 \end{pmatrix}$

Elementary row operations

When solving a system of equations, there are several simple alterations that can be made to the system to produce a new system with the same solutions:

- the order of equations can be switched (the solution doesn't depend on the order)
- an equation of the system can be multiplied by a constant (if $A = B$ then $aA = aB$)
- one equation can be added to, or subtracted from another (if $A = B$ and $C = D$ then $A + C = B + D$).

We would obviously be on the look-out for alterations which produced a simpler system. For example in the case of the postage stamp problem, the system

     $2x + 3y = 21$ became $6x + 9y = 63$ then $6x + 9y = 63$
     $3x + 2y = 19$ $6x + 4y = 38$ $-5y = -25$

In this final simple form, it is easily deduced that $y = 5$, from the second equation. By substitution back into the first equation we get $6x + 45 = 63$, and hence $x = 3$.

When working with the system in the augmented matrix form the three simple alterations become what are known as the *elementary row operations* (EROs):

- two rows can be interchanged
- a row can be multiplied by a constant
- one row can have another row added to it.

EROs can be combined to create more complex row operations, for example you may subtract a multiple of one row from another.

A simple shorthand can be used to record the operations:

R1 $\leftrightarrow$ R2	'interchange row 1 and row 2'
R1 $\rightarrow a \times$ R1	'row 1 becomes a times row 1'
R1 $\rightarrow$ R1 + R2	'row 1 becomes row 1 plus row 2'
R1 $\rightarrow$ R1 – 5 R2	'row 1 becomes row 1 minus 5 times row 2'

Let us re-examine the postage stamp problem using the augmented matrix format:

$$\begin{pmatrix} 2 & 3 & \vdots & 21 \\ 3 & 2 & \vdots & 19 \end{pmatrix}$$

$$\begin{array}{l} R1 \rightarrow 3 \times R1 \\ R2 \rightarrow 2 \times R2 \end{array} \quad \begin{pmatrix} 6 & 9 & \vdots & 63 \\ 6 & 4 & \vdots & 38 \end{pmatrix}$$

$$R2 \rightarrow R2 - R1 \quad \begin{pmatrix} 6 & 9 & \vdots & 63 \\ 0 & -5 & \vdots & -25 \end{pmatrix}$$

At this stage the non-zero entries in the left-hand side of the augmented matrix form a triangle. This is often referred to as the *upper triangular form*. We can re-interpret the matrix as a system of equations: $6x + 9y = 63$ and $-5y = -25$ and proceed to solve as before or we could continue:

$$R2 \rightarrow -\tfrac{1}{5}R2 \quad \begin{pmatrix} 6 & 9 & \vdots & 63 \\ 0 & 1 & \vdots & 5 \end{pmatrix}$$

$$R1 \rightarrow R1 - 9R2 \quad \begin{pmatrix} 6 & 0 & \vdots & 18 \\ 0 & 1 & \vdots & 5 \end{pmatrix}$$

$$R1 \rightarrow \tfrac{1}{6}R1 \quad \begin{pmatrix} 1 & 0 & \vdots & 3 \\ 0 & 1 & \vdots & 5 \end{pmatrix}$$

Re-interpretation now gives us $x = 3$, $y = 5$.

EXERCISE 3

1 (i) Express each of these systems of equations in *augmented matrix* form.
   (ii) Reduce it to *upper triangular form* using row operations.
   (iii) Re-express this simplified form as a system of equations and solve.

a $2x + y = 6$
  $x - y = 1$

b $2x + 3y = -1$
  $x - 2y = -4$

c $4x - 3y = 22$
  $2x + 5y = -2$

d $2x - y = -1$
  $3x - 5y = -4$

2 (i) Express each of these systems of equations in augmented matrix form.

(ii) Reduce it to the form $\begin{pmatrix} 1 & 0 & p \\ 0 & 1 & q \end{pmatrix}$ using row operations.

(iii) Hence solve.

a $3x - 5y = -8$
  $2x - y = -3$

b $4x + 7y = 5$
  $3x - 4y = 13$

c $2x - 5y = -4$
  $5x - 2y = -10$

d $6x - 4y = -36$
  $9x + 2y = -6$

3 Each of the following systems contains an equation which is redundant.

a $2x - y = 1$
  $6x - 3y = 3$

b $x + 3y = 7$
  $4x + 12y = 28$

(i) Express each in augmented matrix form.

(ii) Use row operations to simplify it.

(iii) Comment on what happens to the matrix when redundancy exists.

4 Each of these systems is inconsistent.

a $3x - 2y = 1$
  $6x - 4y = 3$

b $4x + 2y = 5$
  $2x + y = 3$

(i) Express each in augmented matrix form.

(ii) Use row operations to simplify it.

(iii) Comment on what happens to the matrix when inconsistency exists.

5 Relative to a certain set of axes, and using suitable units, the flight of a golf ball can be modelled by an equation of the form $y = ax - bx^2$ where y is the height, x is the distance from the tee, and a and b are constants. It is noted that, when $x = 1$, then $y = 9$ and, when $x = 4$, $y = 12$.

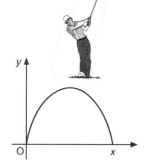

a Form two equations in a and b.

b Solve the system by first expressing it in augmented matrix form.

c How many units of distance from the tee is the ball when it lands ($y = 0$)?

6 In a computer simulation, the routes of four ferries in an estuary have been modelled by the four straight lines:

  Aberline ferry $3x + y = 10$
  Balfour ferry $5x - 2y = -9$
  Clanaig ferry $9x + 3y = 30$
  Dunmuir ferry $6x + 2y = 21$

a Create a suitable augmented matrix to help you find where the Aberline route crosses the Balfour route.

b Attempt to do the same with
  (i) the Aberline/Clanaig routes
  (ii) the Aberline/Dunmuir routes.

c Comment on the geometric interpretation of
  (i) redundancy
  (ii) inconsistency.

3 × 3 systems and matrices

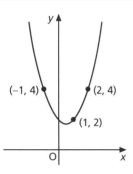

The parabola shown has an equation of the form
$$y = ax^2 + bx + c$$

It passes through the points (−1, 4), (1, 2) and (2, 4). How do we find the constants a, b, c?

Use the points passed through to help you form a system of equations:

(1, 2): $a + b + c = 2$
(2, 4): $4a + 2b + c = 4$
(−1, 4): $a − b + c = 4$

> 3 equations, 3 unknowns: a 3 × 3 system

This system can be reduced to the matrix form $A\mathbf{x} = \mathbf{b}$:

$$\begin{pmatrix} 1 & 1 & 1 \\ 4 & 2 & 1 \\ 1 & -1 & 1 \end{pmatrix} \begin{pmatrix} a \\ b \\ c \end{pmatrix} = \begin{pmatrix} 2 \\ 4 \\ 4 \end{pmatrix}$$

or further reduced to the augmented matrix form

$$\left(\begin{array}{ccc:c} 1 & 1 & 1 & 2 \\ 4 & 2 & 1 & 4 \\ 1 & -1 & 1 & 4 \end{array}\right)$$

> For the sake of reference the entry in the ith row and jth column will be referred to as a_{ij}. As row operations are enacted, the entry will generally change its value.

Row operations will reduce this to upper triangular form.

$$\begin{array}{l} R2 \rightarrow R2 - 4R1 \\ R3 \rightarrow R3 - R1 \end{array} \quad \left(\begin{array}{ccc:c} 1 & 1 & 1 & 2 \\ 0 & -2 & -3 & -4 \\ 0 & -2 & 0 & 2 \end{array}\right) \quad \text{the object being to make } a_{21} = a_{31} = 0$$

$$R2 \leftrightarrow R3 \quad \left(\begin{array}{ccc:c} 1 & 1 & 1 & 2 \\ 0 & -2 & 0 & 2 \\ 0 & -2 & -3 & -4 \end{array}\right) \quad \text{a simple switch which saves some work}$$

$$R3 \rightarrow R3 - R2 \quad \left(\begin{array}{ccc:c} 1 & 1 & 1 & 2 \\ 0 & -2 & 0 & 2 \\ 0 & 0 & -3 & -6 \end{array}\right) \quad \text{making } a_{32} = 0$$

Row 3 gives $-3c = -6$ $\Rightarrow$ $c = 2$
Row 2 gives $-2b = 2$ $\Rightarrow$ $b = -1$
Row 1 gives $a + b + c = 2$ $\Rightarrow$ $a - 1 + 2 = 2$ $\Rightarrow$ $a = 1$

So the equation of the parabola is
$$y = x^2 - x + 2c$$

> In general, back-substitution will be used when dealing with both rows 1 *and* 2.

Gaussian elimination

Gauss

This technique of solving a system of equations by expressing the system in augmented matrix form, reducing it to upper triangular form and then performing back-substitution is called *Gaussian elimination*.

It is named after the famous German mathematician Carl Friedrich Gauss, 1777–1855, who devised it.

Gauss contributed to many branches of mathematics, physics and astronomy. Indeed it was while working on the problem of the orbit of the newly discovered asteroid Ceres that he began to explore numerical analysis and the theory of errors.

EXERCISE 4A

1 For each system of equations:
   (i)   express it as an augmented matrix
   (ii) reduce the matrix, using row operations, to upper triangular form
   (iii) using back-substitution, work out the values of the variables.

   a $x + 2y + z = 8$
   $3x + y - 2z = -1$
   $x + 5y - z = 8$

   b $2x + 3y - z = -1$
   $x - 3y - 2z = 4$
   $5x + y + 3z = 4$

   c $3x + y = 5$
   $x + 2y - 3z = -12$
   $x + 2z = 10$

   d $3x - 4y + z = 24$
   $x - 2y - 2z = 7$
   $x + y + z = 4$

   e $4x + 2y + z = 3$
   $x + 3y + 5z = 3$
   $2x + 3z = 5$

   f $x + y + 5z = 0$
   $4x + y - 6z = -17$
   $x - y - z = 0$

2 A parabola passes through the points (1, 2), (2, 7) and (3, 14).
   It has an equation of the form $y = ax^2 + bx + c$.
   a Use the information to form a 3×3 system of equations.
   b Solve the system by Gaussian elimination.
   c Write down the equation of the parabola.

3 Archaeologists working on an archaeological dig discover the remains of a circular Roman amphitheatre. Using a suitable set of axes and convenient units, the archaeologists positively identify three points on its circumference: (−2, −1), (−1, 2) and (6, 3).
   a Assuming the perimeter has an equation of the form $x^2 + y^2 + 2gx + 2fy + c = 0$, form a system of equations in g, f and c.
   b Use Gaussian elimination to solve the system and identify the equation of the perimeter.
   c What is the radius of the amphitheatre?

4 A set of traffic lights is phased so that Stop periods and the Go periods are separated by Cautionary periods. The pattern runs thus: S – C – G – C – S – C – G – C ...
Each period lasts a fixed length of time, but each type is of different length.
Three observations were made by a traffic warden:
(i)   S – C – G lasted for 185 seconds
(ii) S – C – G – C – S – C – G – C – S lasted 460 seconds
(iii) S – C – G – C – S – C – G lasted 375 seconds.
a Form a set of equations in s, c and g, where these are fittingly named variables, and solve it to find the length of each period.
b At the start of an hour the light turns to Stop. What kind of period will it be in when the hour ends?

Back-substitution using the matrix

Once a matrix has been reduced to upper triangular form, you can continue the row operations until the solution is apparent within the matrix. For example:

$$\begin{pmatrix} 1 & 1 & 1 & \vdots & 2 \\ 0 & -2 & 0 & \vdots & 2 \\ 0 & 0 & -3 & \vdots & -6 \end{pmatrix}$$

Consider the parabola problem cited at the start of Exercise 4A.

$R2 \rightarrow -\frac{1}{2}R2$

$R3 \rightarrow -\frac{1}{3}R3$

$$\begin{pmatrix} 1 & 1 & 1 & \vdots & 2 \\ 0 & 1 & 0 & \vdots & -1 \\ 0 & 0 & 1 & \vdots & 2 \end{pmatrix}$$

Make the *leading* non-zero entry in each row equal to 1 by suitable multiplications.

$R1 \rightarrow R1 - R3$

$$\begin{pmatrix} 1 & 1 & 0 & \vdots & 0 \\ 0 & 1 & 0 & \vdots & -1 \\ 0 & 0 & 1 & \vdots & 2 \end{pmatrix}$$

Perform operations *upwards* to make entries above the leading entries zero.

$R1 \rightarrow R1 - R2$

$$\begin{pmatrix} 1 & 0 & 0 & \vdots & 1 \\ 0 & 1 & 0 & \vdots & -1 \\ 0 & 0 & 1 & \vdots & 2 \end{pmatrix}$$

The solution can now be read off.

EXERCISE 4B

1 For each system:
   (i)   represent it by an augmented matrix
   (ii) reduce this to the form $\begin{pmatrix} 1 & 0 & 0 & \vdots & p \\ 0 & 1 & 0 & \vdots & q \\ 0 & 0 & 1 & \vdots & r \end{pmatrix}$
   (iii) write down the solution to the system of equations.

a $x + 2y + z = 4$
    $2x - y - z = 0$
    $3x + 2y + z = 6$

b $5x - 2y + z = 10$
    $3x - 4y - z = 10$
    $x - 2y - 2z = 3$

c $7x - 2y + 3z = -13$
    $x + 4y + 3z = 11$
    $x + 2y + z = 5$

d $4x + 3y - 2z = 16$
  $x - 2y - 3z = -9$
  $3x - 5y - 2z = -4$

e $x - y - 3z = 1$
  $2x + y - 2z = 9$
  $x - 2y + 2z = 5$

f $y + 3z = 3$
  $2x + 3z = 10$
  $3x + 2y = 0$

2

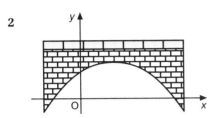

By working with convenient units, the supporting arch of a bridge can be modelled by the equation $y = a + bx - cx^2$. Three points on this arch have been accurately measured as $(1, -3)$, $(2, -28)$ and $(-1, -13)$.

a Use the data to form a system of equations.

b Solve the system by reducing the augmented matrix to the form $\begin{pmatrix} 1 & 0 & 0 & \vdots & p \\ 0 & 1 & 0 & \vdots & q \\ 0 & 0 & 1 & \vdots & r \end{pmatrix}$.

c Write down the equation of the arch.

Redundancy and inconsistency in a 3 × 3 system

Consider the system of equations
$$x + 2y + 2z = 11$$
$$x - y + 3z = 8$$
$$4x - y + 11z = 35$$

$$\begin{pmatrix} 1 & 2 & 2 & \vdots & 11 \\ 1 & -1 & 3 & \vdots & 8 \\ 4 & -1 & 11 & \vdots & 35 \end{pmatrix}$$

R2 → R2 − R1
R3 → R3 − 4R1
$$\begin{pmatrix} 1 & 2 & 2 & \vdots & 11 \\ 0 & -3 & 1 & \vdots & -3 \\ 0 & -9 & 3 & \vdots & -9 \end{pmatrix}$$

R3 → R3 − 3R2
$$\begin{pmatrix} 1 & 2 & 2 & \vdots & 11 \\ 0 & -3 & 1 & \vdots & -3 \\ 0 & 0 & 0 & \vdots & 0 \end{pmatrix}$$

The row of zeros tells us that this equation is redundant, so that there is not a unique solution to the system. There is in fact an infinite number of solutions:

given any z then, from row 2: $y = \dfrac{z + 3}{3}$;

from row 1: $x = 11 - 2z - 2y$

which simplifies to $x = \dfrac{27 - 8z}{3}$

The general solution can be quoted as: $x = \dfrac{27 - 8z}{3}$; $y = \dfrac{z + 3}{3}$; $z = z$

A particular solution can be found by assigning a value to z.
For example, if $z = 3$ the solution would be $x = 1$, $y = 2$, $z = 3$.

Consider the system of equations

$$x + 2y + 2z = 11$$
$$2x - y + z = 8$$
$$3x + y + 3z = 18$$

$$\begin{pmatrix} 1 & 2 & 2 & \vdots & 11 \\ 2 & -1 & 1 & \vdots & 8 \\ 3 & 1 & 3 & \vdots & 18 \end{pmatrix}$$

$$\begin{pmatrix} 1 & 2 & 2 & \vdots & 11 \\ 0 & -5 & -3 & \vdots & -14 \\ 0 & -5 & -3 & \vdots & -15 \end{pmatrix}$$

$$\begin{pmatrix} 1 & 2 & 2 & \vdots & 11 \\ 0 & -5 & -3 & \vdots & -14 \\ 0 & 0 & 0 & \vdots & 1 \end{pmatrix}$$

Row 3 suggests that $0 = 1$, which tells us that the system of equations is in fact inconsistent and that there are no solutions.

EXERCISE 5

1 Attempt to reduce each of the following systems of equations to upper triangular form.
  • Where this is possible quote the unique solution.
  • Where there is a redundant equation, find a general solution.
  • Where there is inconsistency, declare that there are no solutions.

  a $3x + 2y + 5z = 0$
  $2x + y - 2z = 5$
  $7x + 4y + z = 10$

  b $x + y - z = 4$
  $2x - y + 2z = -2$
  $x - 3y - 4z = -1$

  c $2x - y + 3z = 6$
  $x + y + 2z = 7$
  $4x + y + 7z = 9$

  d $2x - 3y + z = 2$
  $x + y - 3z = 7$
  $5x - 2y - z = 14$

  e $x + 2y - z = 3$
  $x + 3z = 5$
  $4x + 2y + 8z = 10$

  f $5x - 3y - z = -12$
  $2x + y + 3z = 3$
  $20x - y + 13z = -9$

2 The system of equations

$$x + 2y - z = 8$$
$$3x + y + 2z = -1$$
$$x + y + kz = -6$$

has no solutions. What is the value of k?

3 Find the value of k that makes the system of equations

$$x + y + z = 1$$
$$2x + 3y - 2z = -1$$
$$x - y + kz = 7$$

have infinitely many solutions.

4 For what values of d and e will the three equations

$$x + 3y - 2z = 8$$
$$2x + y - 3z = 5$$
$$7x - 4y + dz = e$$

have **a** no solution
b infinitely many solutions
c a unique solution?

Automatic processes

The beauty of Gaussian elimination is that, since it is completely systematic, it can be easily programmed into calculators or into spreadsheets. It can be very easily adapted to suit larger systems such as 4×4, 5×5, etc.

Many real-life applications involve large systems. Many real-life applications involve coefficients given correct to, say, three decimal places. In either case, solving such systems by hand becomes impractical.

Solving an $n \times n$ system

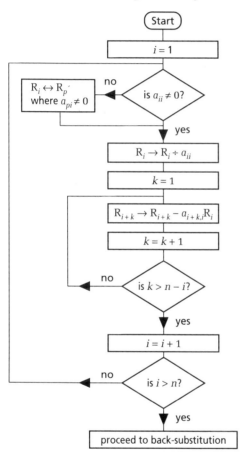

Start at row 1.

Is the entry in the main diagonal non-zero? If not, do a row swap.

Now make this entry = 1 (not actually essential, but handy).

This loop makes all the entries below the main diagonal entry equal to zero.

Next row

Have we done the last row?

This flowchart should help explain how we can take an $n \times n$ matrix and systematically reduce it to upper triangular form, with the added feature that the main diagonal entries are all equal to 1.

In the next exercise some of the examples can be done by hand, but the rest should be tackled by using an appropriate calculator or a spreadsheet.

1 Use Gaussian elimination to solve the following systems, then check your answer, if you can, using a graphics calculator or spreadsheet.

a
$$w + x + y + z = 5$$
$$2w + 3x + y + 2z = 9$$
$$w - x - y + 3z = 1$$
$$2w - x + 2y - z = 4$$

b
$$2w - x + 2y + z = -1$$
$$w + 2x - 3y + 2z = 3$$
$$-w + x - 5y - z = 2$$
$$2w - 2x + 2y + 3z = -5$$

c
$$w + x + y + z = 14$$
$$w - x - y + z = -8$$
$$w - x + y - z = 0$$
$$w - x - y - z = -12$$

d
$$w + x + y = 9$$
$$w + y + z = 7$$
$$w + x + z = 5$$
$$x + y + z = 9$$

2 a The sum of the first n whole numbers, $1 + 2 + 3 + \cdots + n$, can be found using a formula of the form $S_n = an^2 + bn + c$.
From the fact that $S_1 = 1$, $S_2 = 3$ and $S_3 = 6$, form a system of equations and solve it to find the formula.

b The sum of the first n square numbers, $1^2 + 2^2 + 3^2 + \cdots + n^2$, can be found using a formula of the form $S_n = an^3 + bn^2 + cn + d$.
From the fact that $S_1 = 1$, $S_2 = 5$, $S_3 = 14$ and $S_4 = 30$, form a system of equations and solve it to find the formula.

c The sum of the first n cubes, $1^3 + 2^3 + 3^3 + \cdots + n^3$, can be found using a formula of the form $S_n = an^4 + bn^3 + cn^2 + dn$. Find the formula.

3 The rotor blade of a helicopter moves in such a way that the distance, y metres, from its tip to the back of the craft can be calculated using a formula of the form
$$y = a \sin x° + b \cos x° + c$$
where y is the distance, $x°$ is the angle rotated through measured from a certain starting position and a, b and c are constants.
Measuring to three significant figures the following table was constructed.

$x°$	y(m)
30	11.00
100	8.26
120	6.60

a Use the data to form a system of equations. Work to three significant figures.
b Use Gaussian elimination to solve the system and hence construct the formula.
c What is the range of the function $y(x)$?
d Using a calculator, explore the above situation.
How does altering the number of significant figures in the working affect the solution?

Approximate data

When a measurement is made, or rounding takes place, the resulting data takes on a *fuzziness*. For example, when we say $x = 51$ to the nearest whole number we mean $50.5 \leq x < 51.5$. This is often quoted as $x = 51 \pm 0.5$. The 0.5 is called the *absolute error* in the measurement.

51

The size of the error is often quoted as a percentage of the associated measurement. An error of 0.5 in 51 represents approximately a 1% error in the measurement. This *percentage error* is a better way of judging just how bad the error is.

Similarly $y = 50$ to the nearest whole number means $49.5 \leq y < 50.5$.
The sum $x + y$, which should be 101, could be as low as $50.5 + 49.5 = 100$ or as high as $51.5 + 50.5 = 102$. That is, $100 \leq x + y < 102$ or $x + y = 101 \pm 1$.
The sum has an error of 1%, which is acceptable in many situations.

The difference $x - y$, which should be 1, could be as low as $50.5 - 50.5 = 0$ or as high as $51.5 - 49.5 = 2$. That is, $0 \leq x - y < 2$ or $x - y = 1 \pm 1$.
The difference has an error of 100%: the result of such a subtraction has no meaning. When rounded numbers close to each other in value are subtracted, an unacceptable error builds up. This situation is to be avoided!

When working with equations in real-life situations we must watch out for such occasions.

A metal rod, 51 mm measured to the nearest mm, is cooled down and contracts. Its length is then measured as 50 mm, again working to the nearest mm.
By how much has it contracted?
The equation $51 - x = 50$ has to be solved. This means that $x = 51 - 50$, but this is exactly the situation above! $x = 1 \pm 1$ mm, a 100% error.
No useful result can be gleaned from this equation.

EXERCISE 7

1 Each of the following measurements has been rounded. State:
   (i) the range of possible values the measurement can take
   (ii) the absolute error
   (iii) the percentage error.
   a $x = 12$ **b** $y = 25$ **c** $z = 46$
   d $p = 2.4$ **e** $q = 7.7$ **f** $r = 9.1$
   g $a = 0.45$ **h** $b = 0.36$ **i** $c = 0.03$

2 For each pair of values:
   (i)   state the highest and lowest value of the sum and difference $x + y$ and $x - y$
   (ii) state the absolute error in each result
   (iii) express the error as a percentage of the expected answer
   (iv) comment on the errors which you feel are unacceptably large.
   a $x = 28$, $y = 7$ **b** $x = 78$, $y = 77$ **c** $x = 6.4$, $y = 6.3$
   d $x = 0.81$, $y = 0.03$ **e** $x = 1.25$, $y = 1.00$ **f** $x = 0.010$, $y = 0.011$

3 The numbers in the following equations are rounded.

Quote the solutions in the form $x \pm e$, where e is the absolute error in the answer.

a $x - 34 = 12$ **b** $x - 1.4 = 0.9$ **c** $x + 78 = 79$

d $x + 7 = 92$ **e** $x - 1 = 1$ **f** $x + 0.5 = 0.6$

Gaussian elimination on a spreadsheet

For the next section, we wish to be able to study changes to the coefficients in a system of equations and the effect these changes have on the solutions.
Towards this end this spreadsheet has been devised.

	A	B	C	D	E	F	G
1		1	1	1	19	x=	=E21
2		3	16	1	95	y=	=E22
3		2	9	1	59	z=	=E23
4							
5	R1=R1/a11	=B1/$B$1	=C1/$B$1	=D1/$B$1	=E1/$B$1		
6	R2=R2-a21*R1	=B2-$B$2*B5	=C2-$B$2*C5	=D2-$B$2*D5	=E2-$B$2*E5		
7	R3=R3-a31*R1	=B3-$B$3*B5	=C3-$B$3*C5	=D3-$B$3*D5	=E3-$B$3*E5		
8							
9	R1	=B5	=C5	=D5	=E5		
10	R2=R2/a22	=B6/$C$6	=C6/$C$6	=D6/$C$6	=E6/$C$6		
11	R3=R3-a32*R2	=B7-$C$7*B10	=C7-$C$7*C10	=D7-$C$7*D10	=E7-$C$7*E10		
12							
13	R1	=B9	=C9	=D9	=E9		
14	R2	=B10	=C10	=D10	=E10		
15	R3=R3/a33	=B11/$D$11	=C11/$D$11	=D11/$D$11	=E11/$D$11		
16	Going up						
17	R1=R1-a13*R3	=B13-$D$13*B19	=C13-$D$13*C19	=D13-$D$13*D19	=E13-$D$13*E19		
18	R2=R2-a23*R3	=B14-$D$14*B19	=C14-$D$14*C19	=D14-$D$14*D19	=E14-$D$14*E19		
19	R3	=B15	=C15	=D15	=E15		
20							
21	R1=R1-a12*R2	=B17-$C$17*B22	=C17-$C$17*C22	=D17-$C$17*D22	=E17-$C$17*E22		
22	R2	=B18	=C18	=D18	=E18		
23	R3	=B19	=C19	=D19	=E19		

- The augmented matrix occupies cells B1 to E3.
- Column A records the row operations.
- B5 to E7 give the transformed matrix after one sweep of operations.
- B9 to E11 after a second sweep of operations.
- B13 to E15 give the matrix in upper triangular form.
- B17 to E19 give the first step of back-substitution.
- B21 to E23 give the final step of back-substitution.
- Cells F1 to G3 display the solution for convenience beside the original system.

	A	B	C	D	E	F	G
1		1	1	1	19	x=	8.00
2		3	16	1	95	y=	4.00
3		2	9	1	59	z=	7.00
4							
5	R1=R1/a11	1.00	1.00	1.00	19.00		
6	R2=R2-a21*R1	0.00	13.00	-2.00	38.00		
7	R3=R3-a31*R1	0.00	7.00	-1.00	21.00		
8							
9	R1	1.00	1.00	1.00	19.00		
10	R2=R2/a22	0.00	1.00	-0.15	2.92		
11	R3=R3-a32*R2	0.00	0.00	0.08	0.54		
12							
13	R1	1.00	1.00	1.00	19.00		
14	R2	0.00	1.00	-0.15	2.92		
15	R3=R3/a33	0.00	0.00	1.00	7.00		
16	Going up						
17	R1=R1-a13*R3	1.00	1.00	0.00	12.00		
18	R2=R2-a23*R3	0.00	1.00	0.00	4.00		
19	R3	0.00	0.00	1.00	7.00		
20							
21	R1=R1-a12*R2	1.00	0.00	0.00	8.00		
22	R2	0.00	1.00	0.00	4.00		
23	R3	0.00	0.00	1.00	7.00		

The illustrated example shows the solution of the system

$$x + \quad y + z = 19$$
$$3x + 16y + z = 95$$
$$2x + \quad 9y + z = 59$$

Note the final form of the augmented matrix:

$$\begin{pmatrix} 1 & 0 & 0 & 8 \\ 0 & 1 & 0 & 4 \\ 0 & 0 & 1 & 7 \end{pmatrix}$$

Ill-conditioning

Consider the following story.

Two bags of mixed screws are weighed, *to the nearest gram.*
The bags are then sorted out and it is discovered that the situation can be modelled by the system of equations:

| bag 1 | $19x + 18y = 55$ | 19 of one type + 18 of the other weighs 55 g |
| bag 2 | $20x + 19y = 58$ | 20 of one type + 19 of the other weighs 58 g |

where x and y are the weights of one of each type of screw, measured in grams. Solving this system leads to the conclusion that type X weigh 1 g each and type Y weigh 2 g each. However, the 55 g measurement represents 55 ± 0.5 g and the 58 g similarly represents 58 ± 0.5 g, giving percentage errors of around 1%.

Type X   Type Y

The table below shows the variety of conclusions we come to as we explore these ranges.

For example:
$19x + 18y = 54.5$
$20x + 19y = 57.5$
gives $x = 0.5$, $y = 2.5$

	57.5	58	58.5
54.5	(0.5, 2.5)	(−8.5, 12)	(−17.5, 21.5)
55	(10, −7.5)	(1, 2)	(−8, 11.5)
55.5	(19.5, −17.5)	(10.5, −8)	(1.5, 1.5)

From the table we see $-17.5 \leq x \leq 19.5$ or $x = 1 \pm 18.5$ over 1000% error

and $-17.5 \leq y \leq 21.5$ or $x = 2 \pm 19.5$ 975% error.

The actual result is being *swamped* by the error.

Geometrically, when solving a pair of simultaneous equations, we are finding where two lines intersect. If these lines are almost parallel then a slight shift in the position of even one line will result in a big shift in the point of intersection. [Check the gradients of the above *lines*.]

When, like this, a small change in any of the values in a system of equations leads to a disproportionate change in the solutions then the equations are said to be *ill-conditioned.* When this occurs we have no confidence in the results obtained from such a system.

Similar effects occur in larger systems of equations. Imagine a 3×3 system which leads to the above system after row operations, for example:

$$11x + 12y + 3z = 44$$
$$10x + 10y + z = 33$$
$$42x + 43y + 6z = 146$$

As it stands, the solution is $x = 1$, $y = 2$, $z = 3$. Exploring only the effect of *fuzziness* of 44:

altering equation 1 to $11x + 12y + 3z = 43.5$ gives $x = -7.5$, $y = 11$, $z = -2$;

altering equation 1 to $11x + 12y + 3z = 44.5$ gives $x = 9.5$, $y = -7$, $z = 8$;

so *at the very least* we have $x = 1 \pm 8.5$: a percentage error of 850%.

If you have access to a calculator or spreadsheet, explore the parameters of this system.

EXERCISE 8

1 For each of the following systems of equations:
   (i) copy and complete the table
   (ii) identify the range for both x and y-values
   (iii) work out the percentage errors
   (iv) say whether you think the system is ill-conditioned or not.

   a
   $3x + 5y = 17$
   $6x + 11y = 35$

	34.5	35	35.5
16.5			
17			
17.5			

b

$$x + 6y = 13$$
$$2x + y = 4$$

	3.5	4	4.5
12.5			
13			
13.5			

c

$$7x + 4y = 26$$
$$5x + 3y = 19$$

	18.5	19	19.5
25.5			
26			
26.5			

2 Identify which of the following systems you think are ill-conditioned.

a $2x + 9y = 17$
    $3x - 5y = 7$

b $9x + 8y = 1$
    $8x + 7y = 1$

c $8x + 13y = 18$
    $3x + 5y = 7$

d $9x - 10y = 91$
    $8x - 9y = 82$

3 At a cross-roads, traffic lights automatically contol the flow of two streams of traffic. An observer times the changes.

10 turns for the eastbound traffic plus 10 turns for the northbound traffic took 1700 seconds.

10 turns for the eastbound traffic plus 11 turns for the northbound traffic took 1790 seconds.

a Form a system of equations using x and y seconds to represent the time it takes for 1 turn of the eastbound and northbound traffic respectively.

b Solve the system to find a value for x and y.

c Assuming the original timings were made to the nearest second, explore the confidence you can place on your solutions.

4 The spreadsheet on page 134 examines the system of equations.

$$x + y + z = 19$$
$$3x + 16y + z = 95$$
$$2x + 9y + z = 59$$

By considering small changes to the values on the right-hand side of the equations, decide whether or not the system is ill-conditioned.

5 Explore the augmented matrix below where the left-hand side is a 4×4 symmetric matrix called *Wilson*'s matrix.

$$\begin{pmatrix} 10 & 7 & 8 & 7 & 1 \\ 7 & 5 & 6 & 5 & 1 \\ 8 & 6 & 10 & 9 & 1 \\ 7 & 5 & 9 & 8 & 1 \end{pmatrix}$$

a Change just one of the entries in the right-hand column and note the resulting change in the solution of the related system of equations.

b Examine different right-hand columns.

CHAPTER 5 REVIEW

1 Solve this 2×2 system of equations:

   a by elimination

   b by using a matrix.

$$2x + y = 1$$
$$3x - 2y = 5$$

2 **a** Express this 3×3 system of equations as:

     (i) a matrix equation (in the form $Ax = b$)

     (ii) in augmented matrix form.

$$2x + y + z = 2$$
$$3x + 2y - z = 6$$
$$x - y \quad = 0$$

   b Reduce the matrix to upper triangular form.

   c Solve the system with the aid of back-substitution.

3 Solve the following system of equations using Gaussian elimination.

$$x + y + z = 1$$
$$2x - y - z = 14$$
$$x - 2y + z = 4$$

4 In 1804, the first asteroid, Ceres, was discovered. Gauss, using a very small number of observations, was able to compute its elliptical orbit, sparking off *his* interest in numerical analysis and other people's interest in his methods. Using suitable axes, an ellipse has the equation $ax^2 + by^2 + cx - 8y + 1 = 0$. It passes through the points $(3, 1)$, $(-1, 1)$ and $(1, 2)$.

   a Form a 3×3 system of equations using the data.

   b Solve the system using Gaussian elimination.

   c Write down the equation of the ellipse.

5 Examine this system of equations.

$$x + y + z = 2$$
$$x + y \quad = 5$$
$$y + z = -2$$

   a Reduce the associated augmented matrix to the form $\begin{pmatrix} 1 & 0 & 0 & p \\ 0 & 1 & 0 & q \\ 0 & 0 & 1 & r \end{pmatrix}$

   b Hence declare the solution of the system.

6 Under what conditions does the system of equations

$$2x + y + z = 1$$
$$x + 2y + 2z = 1$$
$$3x + y + pz = q$$

   a have no solution

   b have infinitely many solutions?

7 Which system would you consider ill-conditioned?

   (i) $2x + 7y = 5$ **(ii)** $2x - 7y = 5$

      $3x + 10y = 6$ $3x + 10y = 6$

CHAPTER 5 SUMMARY

1 (i) A *matrix* is a table of elements which have been organised in rows and columns.

(ii) The number of rows and columns dictate the order of the matrix.

(iii) The matrix shown here is a 3×2 matrix. $\qquad A = \begin{pmatrix} 2 & 3 \\ 4 & 1 \\ 1 & 2 \end{pmatrix}$

(iv) The matrix has been given the name, A, and the entry in the ith row and jth column is referred to as a_{ij}: $a_{21} = 4$.

2 A system of equations can be represented by a matrix equation of the form $A\mathbf{x} = \mathbf{b}$, for example:

$$\begin{aligned} 2x + 3y + z &= 6 \\ 3x + y + 3z &= 7 \\ x + 2y + z &= 4 \end{aligned} \quad \text{can be represented by} \quad \begin{pmatrix} 2 & 3 & 1 \\ 3 & 1 & 3 \\ 1 & 2 & 1 \end{pmatrix} \begin{pmatrix} x \\ y \\ z \end{pmatrix} = \begin{pmatrix} 6 \\ 7 \\ 4 \end{pmatrix}$$

3 This form can be further compacted into the *augmented matrix* form:

$$\left(\begin{array}{ccc|c} 2 & 3 & 1 & 6 \\ 3 & 1 & 3 & 7 \\ 1 & 2 & 1 & 4 \end{array} \right)$$

4 There are three *elementary row operations* (EROs) which can be performed on a matrix:

(i) two rows can be interchanged.	R1 $\leftrightarrow$ R2
(ii) a row can be multiplied by a constant	R1 $\rightarrow$ aR1
(iii) one row can have another row added to it.	R1 $\rightarrow$ R1 + R2

5 A combination of EROs can produce another row operation. For example, one row can have a multiple of another row added to it: R1 $\rightarrow$ R1 + aR2.

6 (i) If a system of equations has a unique solution then the augmented matrix can be reduced to upper triangular form by row operations.

(ii) If an $n \times n$ system has infinitely many solutions then, when simplified, the augmented matrix will contain at least one row of zero entries.

(iii) If a system has no solutions then at least one row will suggest that $0 = a$ where a is non-zero.

7 When a system has a unique solution, it can be systematically worked out by reducing the augmented matrix to triangular form and then performing back-substitution. This technique is called *Gaussian elimination.*

8 When small changes in the coefficients produce disproportionately large changes in the solution, the system is said to be *ill-conditioned.*

When the data used is rounded data and the system is ill-conditioned, then no confidence can be put on the results obtained.

Answers

CHAPTER 1.1

Exercise 1 (page 1)

1 a Freq. 1, 3, 3, 1 **b** Freq. 1, 4, 6, 4, 1

2 a (i) $x^4 + 4x^3y + 6x^2y^2 + 4xy^3 + y^4$

(ii) $x^5 + 5x^4y + 10x^3y^2 + 10x^2y^3 + 5xy^4 + y^5$

(iii) $x^6 + 6x^5y + 15x^4y^2 + 20x^3y^3 + 15x^2y^4 + 6xy^5 + y^6$

3 a (i) $u_n = \frac{1}{2}n(n-1)$

(ii) $u_n = \frac{1}{6}n(n-1)(n-2)$

b (i) $x^8 + 8x^7y + 28x^6y^2 + 56x^5y^3 + 70x^4y^4 + 56x^3y^5 + 28x^2y^6 + 8xy^7 + y^8$

(ii) $x^9 + 9x^8y + 36x^7y^2 + 84x^6y^3 + 126x^5y^4 + 126x^4y^5 + 84x^3y^6 + 36x^2y^7 + 9xy^8 + y^9$

(iii) $x^{10} + 10x^9y + 45x^8y^2 + 120x^7y^3 + 210x^6y^4 + 252x^5y^5 + 210x^4y^6 + 120x^3y^7 + 45x^2y^8 + 10xy^9 + y^{10}$

c (i) 1, 5, 10, 10, 5, 1; $(n, r) = (n, n-r)$

(ii) No; (a, b) is only defined for $b \le a$.

d (i) $(n, 2) + (n+1, 2)$
$= \frac{1}{2}n(n-1) + \frac{1}{2}n(n+1) = n^2$

Exercise 2A (page 5)

1 a (i) 24 **(ii)** 720 **(iii)** 1

b (i) -24 **(ii)** error **(iii)** error

2 a 24 **b** 5040 **c** 8.07×10^{67}

3 a (i) 840 **(ii)** 5040 **(iii)** 12

b since denominator is shorter than the numerator, all its terms are included in the numerator and so divides.

4 a 10 626 **b** 5040

c (i) 120 **(ii)** 1

5 a (i) 35 **(ii)** 210 **(iii)** 12 **(iv)** 35 **(v)** 210

b proof

c $^7C_4$, $^{10}C_6$, $^{12}C_{11}$, $^7C_3$, $^{10}C_4$

d $^nC_r = \dfrac{n!}{r!(n-r)!}$;

$^nC_{n-r} = \dfrac{n!}{(n-r)!(n-(n-r))!} = \dfrac{n!}{r!(n-r)!}$

6 a (i) 22 100 **(ii)** 635 000 000 000 **(iii)** 1

b 13 983 816

c $\frac{1}{66}$

Exercise 2B (page 7)

1 a 210 **b** 45 **c** 120

2 a 55 **b** 55 **c** $^{11}C_2 = {}^{11}C_9$

3 a (i) 10 **(ii)** 5 **(iii)** 5

b n sides means n vertices. Number of joins $= {}^nC_2$. Number of sides $= n$. So number of diagonals $= {}^nC_2 - n$

c $n = 6$

4 a 4 **b** 10 **c** 8 **d** 16

e 3 **f** 5 **g** 6 **h** 12

5 a $n = 4$ **b** $n = 5$ **c** $n = 7$

d $n = 10$

6 a $n = 6$ **b** $n = 11$ **c** $n = 9$

7 a $n = 7$ **b** $n = 10$ **c** $n = 1$ or 4

d $n = 9$

Exercise 3A (page 9)

1 a $a^5 + 5a^4b + 10a^3b^2 + 10a^2b^3 + 5ab^4 + b^5$

b $1 + 6x + 12x^2 + 8x^3$

c $16 + 96b + 216b^2 + 216b^3 + 81b^4$

d $27a^3 + 54a^2b + 36ab^2 + 8b^3$

e $a^4 - 4a^3b + 6a^2b^2 - 4ab^3 + b^4$

f $1 - 3p + 3p^2 - p^3$

g $81 - 108x + 54x^2 - 12x^3 + x^4$

h $8a^5 - 36a^2b + 54ab^2 - 27b^3$

2 a (i) $x^3 + 3x + \dfrac{3}{x} + \dfrac{1}{x^3}$

(ii) $x^4 + 4x^2 + 6 + \dfrac{4}{x^2} + \dfrac{1}{x^4}$

(iii) $x^5 - 5x^3 + 10x - \dfrac{10}{x} + \dfrac{5}{x^3} - \dfrac{1}{x^5}$

(iv) $x^6 - 6x^4 + 15x^2 - 20 + \dfrac{15}{x^2} - \dfrac{6}{x^4} + \dfrac{1}{x^6}$

b expansions with even powers

3 a $66x^{10}y^2$ **b** $13\,608a^3$ **c** $489\,888x^3y^6$

d $2240x^6$ **e** $-36x^2y^7$ **f** $3840xy^4$

4 a $70x^4y^4$ **b** $720a^3$ **c** second term

d $490x^3$ **e** $1215a^4$ **f** 6

5 a $1 + 3x + 3x^2 + x^3 + 3y + 6xy + 3x^2y + 3y^2 + 3xy^2 + y^3$

b (i) $8 + 12a + 6a^2 + a^3 + 24b + 24ab + 6a^2b + 24b^2 + 12ab^2 + 8b^3$

(ii) $1 - 5x + 10x^2 - 10x^3 + 5x^4 - x^5 + 5y - 20xy + 30x^2y - 20x^3y + 5x^4y + 10y^2 - 30xy^2 + 30x^2y^2 - 10x^3y + 10y^3 - 20xy^3 + 10x^2y^3 + 5y^4 - 5xy^4 + y^5$

(iii) $1 + 4x - 4y + 6x^2 - 12xy + 6y^2 + 4x^3 - 12x^2y + 12xy^2 - 4y^3 + x^4 - 4x^3y + 6x^2y^2 + y^4$

6 Let $x = 1$

7 $4 + 10x + 6x^2 + x^3 = (x + 2)^3 - 2x - 4$

8 a (i) 0.384 **(ii)** 0.512 **b** 0.1536

9 a 0.261 273 6 **b** 0.027 993 6

Exercise 3B (page 11)

1 a 28 **b** 3 **c** 1024

2 $30x^3$; $51x^5$

3 $4x^3$; $4x^{10}$

4 a $1 + 5x + 13x^2 + 22x^3 + 26x^4 + 22x^5 + 13x^6$
$+ 5x^7 + x^8$

b $1 - 5x - 3x^2 + 31x^3 + 19x^4 - 63x^5 - 81x^6$
$- 27x^7$

c $9 - 24x + 46x^2 - 40x^3 + 19x^4 + 20x^5$
$- 26x^6 + 20x^7 + x^8 - 4x^9 + 4x^{10}$

d $1 - 2x - 24x^3 + 48x^4 + 192x^6 - 384x^7$
$- 512x^9 + 1024x^{10}$

5 a $x^5 - x^3 - 2x + \dfrac{2}{x} + \dfrac{1}{x^3} - \dfrac{1}{x^5}$

b $x^6 - 2x^4 - x^2 + 4 - \dfrac{1}{x^2} - \dfrac{2}{x^4} + \dfrac{1}{x^6}$

c when the powers have the same parity

d -40

6 $-3618a^5$, $-7290a^6$

7 $\dfrac{7}{18}$

8 a $^{19}C_r 3^r 2^{19-r}$ $^{19}C_{r+1}3^{r+1}2^{18-r}$

b $r = 11$

9 $u_8 : u_7 = 9 : 14$

10 a 4380.740 937 **b** 11.125

c 1 548 288

d term in x and x^2 are both 372 736 when
$x = 2$; term in x^{12} and x^{11} are both 93 184
when $y = \dfrac{1}{2}$

11 a $^{12}C_3 2^9\left(-\dfrac{2}{3}\right)^3 = -33\,374.81$ (2dp)

b $^{12}C_4 3^{12} = 967\,222\,620$

12 a $^{10}C_5 x^5 = 252x^5$

b $^{11}C_4 2^7 x^4 = 42\,240x^4 = {}^{11}C_3 2^8 x^3$

c the terms in x^n and x^{n+1} will have the
same coefficients

13 $^9C_3\left(\dfrac{1}{2}\right)^6\left(\dfrac{1}{3}\right)^3 = {}^9C_4\left(\dfrac{1}{2}\right)^5\left(\dfrac{1}{3}\right)^4 = \dfrac{7}{144}$

14 $x^2 \rightarrow n^2 2^{n-1}$; $x^3 \rightarrow n(n^2 - 1)2^n/6$

15 $1 - nx + \dfrac{1}{2}n(n + 1)x^2 - \dfrac{1}{6}n(n - 1)(n + 4)x^3$

Exercise 4 (page 13)

1 a 1.05 **b** 1.26 **c** 0.648

d 250 000 **e** 2980 **f** 3020

g 0.0156 **h** 1 570 000 000

2 a $2x\delta x$ **b** $3x^2\delta x$

3 a $1 + 6a + 15a^2$ **b** $-1 + 14a - 84a^2$

c $256 + 1024a^2$

4 Expand brackets separately up to terms in
x^2, then multiply.

5 $f'(x) \approx n.x^{n-1}$

6 a $V_n \approx 10\,000\,(1 - 0.1n)$

b $V_n \approx 5000\,(1 + 0.05n)$

Review (page 14)

1 1

1						
1	1					
1	2	1				
1	3	3	1			
1	4	6	4	1		
1	5	10	10	5	1	
1	6	15	20	15	6	1

2 a $p = 20$; $q = 5$; $r = 15$

b $s = 14$; $t = 6$

c (i) 38 760 **(ii)** 167 960

3 a $n = 5$ **b** $p = 8$; $q = 4$

4 a $x^5 - 20x^4 + 160x^3 - 640x^2 + 1280x - 1024$

b $8y^3 - 36y^2 + 54y - 27$

5 a $120x^7$ **b** $525y^3$

6 $^{10}C_5 x^5\left(\dfrac{-2}{x}\right)^5 = -8064$

7 a $2^{11} = 2048$

b (i) 1024 **(ii)** 1024

CHAPTER 1.2

Exercise 1 (page 17)

1 $x + 1 + \dfrac{3}{x + 2}$ **2** $x - 5 + \dfrac{19}{x + 3}$

3 $x + 5 + \dfrac{5}{x - 2}$ **4** $3x + 7 + \dfrac{29}{x - 4}$

5 $3 + \dfrac{2 - 7x}{x^2 + x + 1}$ **6** $1 + \dfrac{3 - 2x}{x^2 + x - 2}$

7 $1 + \dfrac{x - 2}{x^2 - x + 2}$ **8** $x + 3 + \dfrac{3x - 8}{x^2 + 1}$

9 $x^2 - 2x + 4 - \dfrac{7}{x + 2}$

10 $3x^2 + 12x + 46 + \dfrac{188}{x - 4}$

11 $1 + \dfrac{7}{x^2 - 4}$ **12** $x^2 + x - \dfrac{3}{x^2 + 2x}$

13 $x^2 + 1 + \dfrac{x - x^2}{x^3 - x + 1}$ **14** $3x - \dfrac{2x + 1}{x^2 + 3}$

Exercise 2 (page 18)

1 $\dfrac{1}{x - 1} + \dfrac{1}{x + 1}$ **2** $\dfrac{4}{x - 2} + \dfrac{6}{x + 3}$

3 $\dfrac{-1}{x + 1} + \dfrac{5}{x + 5}$ **4** $\dfrac{4}{x - 3} - \dfrac{4}{x + 2}$

5 $\dfrac{1}{x - 1} - \dfrac{1}{x + 2}$ **6** $\dfrac{5}{x + 2} - \dfrac{5}{x + 3}$

7 $\dfrac{4}{x - 2} + \dfrac{1}{x - 3}$ **8** $\dfrac{2}{x + 2} + \dfrac{1}{x - 2}$

Answers

141

9 $\dfrac{4}{2x-1} - \dfrac{3}{x+3}$

10 $\dfrac{3}{x-1} + \dfrac{4}{x+3}$

16 $\dfrac{8}{x-1} - \dfrac{1}{x} - \dfrac{1}{x^2}$

11 $\dfrac{-5}{2(x-3)} + \dfrac{3}{2(x+1)}$

12 $\dfrac{11}{x+3} - \dfrac{5}{x}$

17 $\dfrac{2}{x+1} - \dfrac{1}{x-1} - \dfrac{5}{(x-1)^2}$

13 $\dfrac{2}{x} + \dfrac{1}{x+2}$

14 $\dfrac{-2}{x} - \dfrac{1}{x+6}$

18 $\dfrac{3}{x} - \dfrac{1}{x+1} + \dfrac{2}{(x+1)^2}$

15 $\dfrac{1}{x} - \dfrac{7}{x+4}$

16 $\dfrac{4}{5-x} - \dfrac{3}{x+1}$

Exercise 4 (page 20)

17 $\dfrac{1}{3x+1} + \dfrac{3}{2x+3}$

18 $\dfrac{2}{2x-3} - \dfrac{3}{4x-1}$

1 $\dfrac{1}{x+1} + \dfrac{2-x}{x^2-x+1}$

19 $\dfrac{2}{x-5} + \dfrac{3}{x+6}$

20 $\dfrac{3}{x-1} + \dfrac{2}{x-3}$

2 $\dfrac{1}{x-1} + \dfrac{x+3}{x^2+2x+5}$

21 $\dfrac{1}{x+1} + \dfrac{3}{x+2}$

22 $\dfrac{10}{3x} + \dfrac{2}{3(x-3)}$

3 $\dfrac{9}{x-3} + \dfrac{3-2x}{2x^2-x+1}$

23 $\dfrac{1}{2x-1} + \dfrac{2}{2x+1}$

24 $\dfrac{3}{x+1} - \dfrac{2}{2x+1}$

4 $\dfrac{2}{x-1} + \dfrac{1}{x^2+x+1}$

25 $\dfrac{1}{x+1} + \dfrac{2}{x+2} - \dfrac{1}{x-1}$

5 $\dfrac{1}{x-33} + \dfrac{x+4}{x^2-x+1}$

26 $\dfrac{1}{x} + \dfrac{2}{x-1} + \dfrac{3}{x-2}$

6 $\dfrac{x+3}{x^2+2} - \dfrac{1}{x}$

Exercise 3 (page 19)

7 $\dfrac{3}{x-1} - \dfrac{2x}{x^2+3}$

1 $\dfrac{3}{(x-1)^2} + \dfrac{8}{x-1} - \dfrac{5}{x-2}$

8 $\dfrac{4}{x+1} + \dfrac{1}{x^2+x+3}$

2 $\dfrac{2}{x-3} + \dfrac{4}{x+2} - \dfrac{3}{(x+2)^2}$

9 $\dfrac{3}{x-2} + \dfrac{x+1}{x^2-x+1}$

3 $\dfrac{3}{x+1} + \dfrac{1}{x-2} + \dfrac{5}{(x-2)^2}$

10 $\dfrac{1}{2x+1} + \dfrac{1}{x^2+x+2}$

4 $\dfrac{1}{x^2} + \dfrac{2}{x} - \dfrac{1}{x-1}$

11 $\dfrac{2}{x-1} + \dfrac{2-x}{x^2-x+1}$

5 $\dfrac{4}{(x-1)^2} + \dfrac{4}{x-1} - \dfrac{3}{x}$

12 $\dfrac{6}{x+2} + \dfrac{2x-3}{x^2-x+2}$

6 $\dfrac{4}{x-1} + \dfrac{3}{(x+2)^2}$

13 $\dfrac{1}{3(x-1)} + \dfrac{2x-5}{3(x^2+x+1)}$

7 $\dfrac{7}{9(1-x)} + \dfrac{7}{9(x+2)} + \dfrac{1}{3(x+2)^2}$

14 $\dfrac{3x}{x^2+2x+4} - \dfrac{2}{x-2}$

8 $\dfrac{1}{2-x} + \dfrac{2}{2x+1} + \dfrac{10}{(2x+1)^2}$

15 $\dfrac{1}{x-1} + \dfrac{3x-1}{x^2+x+1}$

9 $\dfrac{1}{x} - \dfrac{2}{2x+1} + \dfrac{3}{(2x+1)^2}$

10 $\dfrac{1}{x} - \dfrac{3}{3x-2} + \dfrac{1}{x^2}$

Exercise 5 (page 21)

11 $\dfrac{9}{1-3x} + \dfrac{3}{x} + \dfrac{1}{x^2}$

1 $\dfrac{1}{3(1-x)} - \dfrac{2}{3(2+x)}$

12 $\dfrac{1}{x+1} - \dfrac{1}{x-3} + \dfrac{4}{(x-3)^2}$

2 $\dfrac{4}{7(2x+1)} + \dfrac{5}{7(x-3)}$

13 $\dfrac{1}{x-2} - \dfrac{2}{x} + \dfrac{3}{x^2}$

3 $\dfrac{2}{x-2} + \dfrac{1}{x+1}$

14 $\dfrac{3}{2x+1} + \dfrac{2}{(x-5)^2}$

4 $\dfrac{1}{2(x+1)} - \dfrac{1}{2(x-1)} + \dfrac{1}{(x-1)^2}$

15 $\dfrac{4}{x+5} - \dfrac{1}{x-1} - \dfrac{2}{(x-1)^2}$

5 $\dfrac{7x}{x^2+1} - \dfrac{4}{x}$

6 $\dfrac{3}{4x} - \dfrac{3}{4(x-2)} + \dfrac{3}{2(x-2)^2}$

7 $\dfrac{1}{4x} - \dfrac{x}{4(x^2+4)}$

8 $\dfrac{7}{x+1} - \dfrac{7}{x} + \dfrac{7}{x^2} - \dfrac{3}{x^3}$

9 $\dfrac{12}{5(x-3)^2} + \dfrac{13}{25(x-3)} - \dfrac{13}{25(x+2)}$

10 $\dfrac{8}{7(x+2)} + \dfrac{13x-12}{7(x^2+3)}$

11 $\dfrac{1}{1-x} + \dfrac{x+2}{1+x+x^2}$

12 $\dfrac{1}{x-1} - \dfrac{1}{x} + \dfrac{1}{x+1}$

13 $\dfrac{1}{5(x-2)} + \dfrac{1}{2(x+1)} - \dfrac{7}{10(x+3)}$

14 $\dfrac{7}{16(x-2)} + \dfrac{9}{16(x+2)} - \dfrac{1}{x} + \dfrac{1}{4x^2}$

15 $\dfrac{\sqrt{2}}{4(x-\sqrt{2})} - \dfrac{\sqrt{2}}{4(x+\sqrt{2})}$

16 $\dfrac{1}{x-3} + \dfrac{6}{(x-3)^2} + \dfrac{9}{(x-3)^3}$

17 $\dfrac{1}{x+5} + \dfrac{32}{(x-3)^2}$

18 $\dfrac{1}{x+1} - \dfrac{x}{x^2-x+1}$

19 $\dfrac{1}{16(x-2)} + \dfrac{1}{16(x-2)} - \dfrac{x}{8(x^2+4)}$

20 $\dfrac{11}{64(x-3)} - \dfrac{11}{64(x+5)} + \dfrac{5}{8(x-3)^2}$

21 $\dfrac{3(\sqrt{3}-1)}{4(\sqrt{3}-x)} + \dfrac{3(\sqrt{3}+1)}{4(\sqrt{3}+x)} - \dfrac{3}{2(x+1)}$

22 $\dfrac{1}{2(x-1)} + \dfrac{7}{6(x+1)} - \dfrac{1}{6(x-2)} - \dfrac{3}{2(x+2)}$

23 $\dfrac{8}{1-2x} + \dfrac{4}{x} + \dfrac{2}{x^2} + \dfrac{1}{x^3}$

24 $\dfrac{4}{5(x-1)} - \dfrac{1}{(x-1)^2} + \dfrac{1-4x}{5(x^2+4)}$

25 $\dfrac{1}{x} - \dfrac{1}{2(x-1)} + \dfrac{1}{4(x-1)^2} - \dfrac{1}{2(x+1)}$
$\quad - \dfrac{1}{4(x+1)^2}$

26 $\dfrac{1}{16x} - \dfrac{x}{16(x^2+4)} - \dfrac{x}{4(x^2+4)^2}$

Exercise 6 (page 22)

1 a $1 - \dfrac{4}{x+2} + \dfrac{2}{x-1}$

b $x + 1 - \dfrac{1}{x-3} + \dfrac{1}{x+1}$

c $x + \dfrac{3}{x-3} + \dfrac{2}{x-2}$

d $2 + \dfrac{1}{4(x-2)} - \dfrac{1}{4(x+2)}$

e $x + 1 - \dfrac{2}{3(x+1)} + \dfrac{2}{3(x-2)}$

f $1 + \dfrac{2}{x-1} + \dfrac{1}{(x-1)^2}$

g $1 - \dfrac{2}{3x} + \dfrac{2-3\sqrt{3}}{6(x+\sqrt{3})} + \dfrac{2+3\sqrt{3}}{6(x-\sqrt{3})}$

h $x + \dfrac{1}{2x} - \dfrac{5x}{2(x^2+2)}$

i $3x + 1 - \dfrac{1}{x^2} + \dfrac{3}{x} - \dfrac{2}{x+1}$

j $1 + \dfrac{3}{2(x+1)} - \dfrac{3}{2(x-1)}$

k $1 + \dfrac{3}{x+2} - \dfrac{4}{x+1}$

l $x + 6 + \dfrac{4}{x-1} + \dfrac{12}{x-3}$

2 $\dfrac{x^2}{(x+a)(x+b)} =$
$1 - \dfrac{a^2}{(a-b)(x+a)} + \dfrac{b^2}{(a-b)(x+b)};$
$A = 1, B = a^2, C = b^2$

Review (page 23)

1 a $\dfrac{3}{x-4} - \dfrac{2}{x+2}$

b $\dfrac{1}{(x+1)^2} - \dfrac{2}{x+1} - \dfrac{1}{x-2}$

c $\dfrac{3}{x-1} + \dfrac{2x+2}{x^2+2x+3}$

2 a $-\dfrac{3}{x} + \dfrac{2}{x-1} + \dfrac{1}{x+1}$

b $\dfrac{3}{x^2} - \dfrac{1}{x-1}$

c $\dfrac{1}{x} - \dfrac{2x+3}{x^2+x+1}$

3 a $1 + \dfrac{1}{x-1} - \dfrac{2}{x+2}$

b $x + \dfrac{1}{x} - \dfrac{2}{x^2+1}$

CHAPTER 2.1

Exercise 1A (page 29)

1 5   **2** 12   **3** $6x$ **4** $10x$

5 $3x^2$ **6** $12x^2 + 3$ **7** $-\dfrac{2}{x^2}$ **8** $-\dfrac{8}{x^3}$

Exercise 1B (page 29)

1 $2\cos 2x$ **2** $3\cos 3x$ **3** $2(x-5)$

4 $4(2x+3)$ **5** $-\dfrac{2}{(2x+1)^2}$

6 $-\dfrac{2}{(x-3)^3}$　　**7** $-\sin x$

8 $-2\sin(2x-1)$

b $\dfrac{\sin x \cos(\cos x)}{\sin^2(\cos x)}$　　**c** $-\dfrac{3\cos(3x+2)}{2(\sin(3x+2))^{\frac{3}{2}}}$

4 **a** $\sin x + (\sin x)^{-2} - 1$

b $\cos x\left(1 - \dfrac{2}{\sin^3 x}\right)$

Exercise 2 (page 30)

1 **a** $3x^2 - \sin x$　　**b** $3\cos 3x + 5x^4$

2 **a** $-2x^{-3} + 2x$　　**b** $-\dfrac{8}{x^5} + 10x$

　　c $-\dfrac{9}{x^4} - 2\sin 2x$　　**d** $4\cos 4x - 4\sin 4x$

3 **a** $1 + \dfrac{3}{x^2}$　　**b** $-\dfrac{6}{x^4} - \dfrac{12}{x^5}$

4 **a** 3　　**b** 4

5 **a** 8　　**b** $-\dfrac{1}{2}\sin\dfrac{1}{2} + \dfrac{9}{2}$

Exercise 3A (page 32)

1 **a** $15(5x+1)^2$　　**b** $24(4x-7)^5$

　　c $3(8x-7)(4x^2-7x)^2$　　**d** $2x\cos(x^2)$

2 **a** $-2\sin 2x$

　　b $6(1-7x)(5+2x-7x^2)^2$

　　c $-(2x+3)\sin(x^2+3x)$

　　d $3\sin^2 x \cos x$

3 **a** $\dfrac{24}{(1-4x)^3}$　　**b** $\dfrac{24x}{(1-4x^2)^2}$

　　c $4(2x+3)(x^2+3x+1)^3$

4 **a** $\dfrac{\pi}{180}\cos x°$　　**b** $-\dfrac{\pi}{180}\sin x°$

　　c $\dfrac{\pi}{60}\cos(3x-30)°$

5 **a** $-\dfrac{\cos x}{\sin^2 x}$　　**b** $+\dfrac{\sin x}{\cos^2 x}$

6 **a** $-\sin x \cos(\cos x)$　　**b** $-\cos x \sin(\sin x)$

　　c $\sin x \sin(\cos x)$　　**d** $\cos x \cos(\sin x)$

7 **a** $\dfrac{dy}{dx} = \cos(\sin^{-1}(x))\dfrac{d}{dx}(\sin^{-1}(x))$

　　b $\sqrt{1-x^2}$　　**c** $\dfrac{1}{\sqrt{1-x^2}}$

　　d $-\dfrac{1}{\sqrt{1-x^2}}$

Exercise 3B (page 33)

1 **a** $6\sin 3x \cos 3x$

　　b $-\cos x \sin(2\sin x)$

　　c $2(x+\sin 3x)(1+3\cos 3x)$

　　d $-\sin 2x \sin(\sin^2 x)$

2 **a** $-6\sin(2x+4)\cos^2(2x+4)$

　　b $-\dfrac{6\cos(3x+1)}{\sin^3(3x+1)}$

　　c $\dfrac{2(x+1)}{(x^2+2x+1)^2}\sin\left(\dfrac{1}{x^2+2x+1}\right)$

3 **a** $\dfrac{(2x+1)\sin(x^2+x)}{\cos^2(x^2+x)}$

Exercise 4A (page 35)

1 **a** $x^2(3\sin x + x\cos x)$

　　b $(x+1)(2\cos x - (x+1)\sin x)$

　　c $(7x-1)(x+1)^3(x-1)^2$

　　d $(7x+19)(x+1)(x+7)^4$

　　e $(x-1)^3(4\cos x - (x-1)\sin x)$

2 **a** 0

　　b $\pi - 2$

3 **a** $(2x+1)(4\cos x - (2x+1)\sin x)$

　　b $2\cos 2x \cos 3x - 3\sin 2x \sin 3x$

　　c $4x^3 \sin 3x + 3x^4 \cos 3x$

　　d $4(2x+1)(3x-1)^3(9x+2)$

　　e $(2x+1)\sin 2x + 2x(x+1)\cos 2x$

　　f $x(x-1)(5x^2+5x+2)$

　　g $2\cos 2x \sin 3x + 3\sin 2x \cos 3x$

　　h $x^3(7x+12)(x+3)^2$

　　i $3x^4(2x+5)$

　　j $\cos 2x$

Exercise 4B (page 36)

1 **a** $-150x(1-25x^2)^2$

　　b $-2\sin 2x(6\cos^2 2x - 1)$

　　c $x(x-1)^3(2(3x-1)\cos x - x(x-1)\sin x)$

　　d $4(2x+3)(x^2+3x+1)^3$

2 $\dfrac{1}{4}\left(1 + \dfrac{\pi}{\sqrt{3}}\right)$

3 $\sin\left(\dfrac{\pi^2}{9}\right) + 2\left(\dfrac{\pi^2}{9}\right)\cos\left(\dfrac{\pi^2}{9}\right)$

4 $(16t-5)(9t^2+4) + 18t^2(8t-5)$

Exercise 5A (page 37)

1 **a** $\dfrac{x^3(3x+4)}{(x+1)^2}$　　**b** $\sec^2 x$

　　c $\dfrac{2-2x-x^2}{(x^2+2)^2}$　　**d** $-\dfrac{x\sin x + \cos x}{x^2}$

2 **a** $\dfrac{\sin x - x\cos x}{\sin^2 x}$　　**b** $\dfrac{3(x-6)}{2(x-3)^{\frac{3}{2}}}$

　　c $\dfrac{6-x}{6x^2\sqrt{x-3}}$　　**d** $\dfrac{x(x+16)}{2(x+4)^{\frac{5}{2}}}$

3 **a** $-\csc x \cot x$　　**b** $\sec x \tan x$

　　c $-\csc^2 x$

4 0.5

6 proof

5 2

7 (0, 4)

Exercise 5B (page 38)

1 $\dfrac{2(3x + 2)(3x - 5)}{(2x - 1)^2}$

2 $-\dfrac{1}{2}\cot x \operatorname{cosec} x$

3 $-3\sqrt{2}$

Exercise 6 (page 38)

1 a $\dfrac{2 - 2x - x^2}{(x^2 + 2)^2}$ **b** $-2\cot 2x \operatorname{cosec} 2x$

c $3\sin 6x$ **d** $3(x + 1)^2(2x^3 + x^2 + 1)$

2 $\cos(\sin x)\cos x$

3 $2(x + 1)(\cos 2x - (x + 1)\sin 2x)$

4 $-\dfrac{1}{2}$

5 $\dfrac{(2x + 3)^2(2x^2 + 6x - 15)}{(x^2 + 3x - 1)^2}$

Exercise 7 (page 40)

2 a $2\sec 2x \tan 2x$ **b** $3\sec^2(3x)$

c $-a\operatorname{cosec}(ax)\cot(ax)$

d $-2\operatorname{cosec}(2x + 3)\cot(2x + 3)$

e $-6x\sec x(4 - 3x^2)\tan(4 - 3x^2)$

f $-5\operatorname{cosec}^2(5x)$ **g** $-2x\operatorname{cosec}^2(x^2)$

h $-17\sec^2(1 - 17x)$

3 a $\sec x(2\sec^2 x - 1)$

b $-\sec^2 x \operatorname{cosec}^2(\tan x)$

c $-\operatorname{cosec}(\sin x)\cot(\sin x)\cos x$

d $-6\operatorname{cosec}^2(3x)\cot(3x)$

e $2\sec^2 x \tan x$

f $8\tan 4x \sec^2 4x$

g $\dfrac{1}{2}\tan x \sqrt{\sec x}$

h $\dfrac{1}{2}(1 + \operatorname{cosec} x)^{-\frac{3}{2}}\operatorname{cosec} x \cot x$

4 a $\dfrac{(2x + 1)(1 + \cot x) + (x^2 + x)\operatorname{cosec}^2 x}{(1 + \cot x)^2}$

b $\dfrac{2\sec x(1 + \operatorname{cosec}^2 x)}{(\cot x - \sec x)^2}$

c $\dfrac{(x + 1)(\sec x \tan x - \operatorname{cosec}^2 x) - 2(\sec x + \cot x)}{(x + 1)^3}$

Exercise 8A (page 43)

1 a $5e^{5x}$ **b** $\dfrac{1}{x + 5}$ **c** $14xe^{7x^2}$

d $\dfrac{6}{x}$ **e** $\cos x e^{\sin x}$

2 a $\dfrac{1}{x^2}e^{-\frac{1}{x}}$ **b** $\dfrac{12x^2}{4x^3 - 1}$

c $2\operatorname{cosec} 2x$ **d** $\dfrac{1}{x}\sec^2(\ln x)$

e 1

3 a $2\cot 2x$ **b** 3

c $\dfrac{1}{x\ln x}$ **d** $3\sec^2 3x e^{\tan 3x}$

e $e^{e^x + x}$

4 a $2(x + 3)e^{x+2}$ **b** $3(1 + x)e^x$

c $e^{2x}(2\sin x + \cos x)$ **d** $e^x\left(\ln x + \dfrac{1}{x}\right)$

e $\ln x + 1$

5 a $\dfrac{\ln x - 1}{(\ln x)^2}$ **b** $\dfrac{e^{2x}(2x - 1)}{2x^2}$

c $\dfrac{3(3x + 2 - 3x\ln x)}{x(3x + 2)^2}$ **d** $\dfrac{2 - x - x^2}{e^x}$

e $\dfrac{3 - 2(3x - 1)\ln(3x - 1)}{e^{2x-4}(3x - 1)}$

6 a $2e^{2x}(\cos(2x + 1) - \sin(2x + 1))$

b $e^x \ln x\left(\dfrac{2}{x} + \ln x\right)$

c $\dfrac{1 - \ln x}{2x^2}\cos\left(\dfrac{\ln x}{2x}\right)$

d $\dfrac{e^{\sin x}\cos x(\sin x - 1)}{\sin^2 x}$

e $-\dfrac{1}{x^3}(x\tan x + 2\ln(\cos x))$

Exercise 8B (page 44)

1 c $3^x \ln 3$

d **(i)** $4^x \ln 4$ **(ii)** $6^x \ln 6$ **(iii)** $a^x \ln a$

(iv) $2^{4(x + 1)}\ln 2$

2 a $x = 10^y$ **b** $\ln x = y\ln 10$

c $y = \dfrac{\ln x}{\ln 10}$ **d** $\dfrac{1}{x\ln 10}$

3 a $\dfrac{1}{x\ln 2}$ **b** $\dfrac{1}{x\ln 5}$

c $\dfrac{3}{(3x + 5)\ln 10}$

Exercise 9A (page 46)

1 a **(i)** $f''(x) = 2$ **(ii)** $f'''(x) = 6$

(iii) $f^{iv}(x) = 24$ **(iv)** $f^v(x) = 120$

b $n!$

2 a $f'(x) = 6(2x + 1)^2$; $f''(x) = 24(2x + 1)$

b 4

3 f', f'', f''', f^{iv}

4 a $-2\sin 2x$; $-4\cos 2x$; $8\sin 2x$;

$(-1)^{\left(\frac{2n + 1 - (-1)^n}{4}\right)}.2^n \sin\left(\dfrac{2x + (1 + (-1)^n)}{4}\pi\right)$

b $-x^{-2}, 2x^{-3}, -6x^{-4}, (-1)^n n! x^{-n-1}$

c $x^{-1}, -x^{-2}, 2x^{-3}, (-1)^{n-1}(n-1)! x^{-n}$

d $4e^{4x}, 16e^{4x}, 64e^{4x}, 4^n e^{4x}$

e $\frac{1}{2}x^{-\frac{1}{2}}, \frac{1}{2}(-\frac{1}{2})x^{-\frac{3}{2}}, \frac{1}{2}(-\frac{1}{2})(-\frac{3}{2})x^{-\frac{5}{2}};$

$(-1)^{n-1}\dfrac{1.3.5...(2n-3)}{2^n}x^{-\frac{2n-1}{2}}$

f $e^x + xe^x, 2e^x + xe^x, 3e^x + xe^x, ne^x + xe^x$

5 a $\sec^2 x, 2\sec^2 x \tan x$ **b** $-\tan x, -\sec^2 x$

c $\dfrac{d^n}{dx^n}(\ln(\cos x)) = -\dfrac{d^{n-1}}{dx^{n-1}}(\tan x)$

6 $a^n e^{ax}$

7 $(2x+1)^{-2}, -4(2x+1)^{-3}$

Exercise 9B (page 46)

1 a $\frac{4}{3}(x-1)^{\frac{1}{3}}, \frac{4}{9}(x-1)^{-\frac{2}{3}}$

2 $\frac{3}{2}x^{\frac{1}{2}}, \frac{3}{4}x^{-\frac{1}{2}}, 2$

3 $\ln x, \dfrac{1}{x}, 2$

Review (page 48)

1 $8x - 3$

2 $6x^2 + \dfrac{3}{2x^2}$

3 $\sqrt{3}$

4 $\dfrac{\sqrt{x}}{2}(3\sin x + 2x\cos x)$

5 $-\dfrac{(3x-2)\sin x + 6\cos x}{(3x-2)^3}$

6 $-3\csc^2 3x$

7 $\sec x(2\sec^2 x - 1)$

CHAPTER 2.2

Exercise 1 (page 51)

1 a (i) $v = 8t - 6, a = 8$

  (ii) $v = 10\,\text{m/s}, a = 8\,\text{m/s}^2$

b (i) $v = (t+1)^{-2}, a = -2(t+1)^{-3}$

  (ii) $v = \frac{1}{9}\text{m/s}, a = -\frac{2}{27}\text{m/s}^2$

c (i) $v = (3t+2)^{-\frac{2}{3}}, a = -2(3t+2)^{-\frac{5}{3}}$

  (ii) $v = \frac{1}{4}\,\text{m/s}, a = -\frac{1}{16}\,\text{m/s}^2$

d (i) $v = 450\cos 15t, a = -6750\sin 15t$

  (ii) $v = 69.4\,\text{m/s}, a = 6669\,\text{m/s}^2$

e (i) $v = 1 - t^{-2}, a = 2t^{-3}$

  (ii) $v = \frac{3}{4}\,\text{m/s}, a = \frac{1}{4}\,\text{m/s}^2$

f (i) $v = 2t - 2t^{-3}, a = 2 + 6t^{-4}$

  (ii) $v = 3\frac{3}{4}\,\text{m/s}, a = 2\frac{3}{8}\,\text{m/s}^2$

2 a $225\,\text{m}, -32\,\text{m/s}^2$

b $3\,\text{m}, -6\,\text{m/s}^2; -1\,\text{m}, 6\,\text{m/s}^2$

c $15\,\text{m}, 5\,\text{m/s}^2$

d $10\,\text{m}, -1440\,\text{m/s}^2$

e $8\,\text{m}, 2\,\text{m/s}^2$

f $(10\ln 10 - 10)\,\text{m}, -10\,\text{m/s}^2$

3 a $a = -\frac{1}{4}(t+1)^{-\frac{3}{2}}$

b $a = -\frac{1}{4}v^3$

4 a (i) $1406.25\,\text{m}$ **(ii)** $9.375\,\text{s}$

b (i) $-300\,\text{m/s}$

  (ii) $a = -32\,\text{m/s}^2$ which is constant throughout the motion

5 a $192\,\text{m}$

b $72\,\text{m/s}$

c (i) $18\,\text{m/s}^2;$ **(ii)** $9\,\text{m/s}^2$

6 a $v = \dfrac{4\pi}{3}\,\text{m/s}, a = 0\,\text{m/s}^2$

b (i) $v = 0\,\text{m/s}$ at A and B

  (ii) AB $= 16\,\text{m}$

7 a (i) $10\,\text{m}$ above the ground

  (ii) $v = \dfrac{\pi}{6}\,\text{m/s}, a = 0\,\text{m/s}^2$

b $30\,\text{s}$

c $a = -\dfrac{\pi^2}{360}\,\text{m/s}^2$

8 a $v = -\dfrac{\pi}{4}\,\text{m/h}$

b (i) 1 pm, 7 pm **(ii)** $-\dfrac{\pi^2}{12}\,\text{m/s}^2, \dfrac{\pi^2}{12}\,\text{m/s}^2$

9 a $t = 0\,\text{s}, 2\,\text{s}, 4\,\text{s}$

b $-\dfrac{\pi}{4}\left(1 + \dfrac{1}{\sqrt{21}}\right)\text{m/s}$

c (i) $\frac{3}{4}\pi^2\,\text{m/s}^2$ **(ii)** $0\,\text{m/s}^2$

10 $h = \dfrac{v_1^2}{2g}$

11 a $v = \dfrac{1}{2} - \dfrac{2t}{(1+t^2)^2}\,\text{m/s}$

b $0.485\,\text{m/s}$

c $0.5\,\text{m/s}$

12 a 58.5 units

b (i) $v = \dfrac{c}{t+1}$ **(ii)** $a = -\dfrac{c}{(t+1)^2}$

c 13.1 units/s

Exercise 2 (page 56)

1 a A: endpoint max; B: endpoint; C: horizontal point of inflexion; D: local minimum; E: local maximum

b P: endpoint; Q: local maximum; R: local minimum; S: endpoint; T: endpoint maximum

2 **a** **(i)** $x = 6$

    (ii) $x = 0$: no left deriv.; $x = 2$: left deriv. $\neq$ right deriv.; $x = 4$: left deriv. $\neq$ right deriv.; $x = 6$: no right deriv.

    (iii) $(0, -2)$ endpoint max; $(7, -3)$ endpoint min

  b **(i)** $x = 0$

    (ii) $x = -3$ no left deriv.; $x = -2$ left deriv. $\neq$ right deriv.; $x = -1$ left deriv. $\neq$ right deriv.; $x = 1$ left deriv. $\neq$ right deriv.; $x = 2$: no right deriv.

    (iii) $(2, 3)$ endpoint max; $(-3, -2)$ endpoint min

  c **(i)** $x = \pi, \dfrac{3\pi}{2}$

    (ii) $x = 0$ no left deriv.; $x = \dfrac{\pi}{4}$ left deriv. $\neq$ right deriv.; $x = \dfrac{5\pi}{4}$ left deriv. $\neq$ right deriv.; $x = 2\pi$ no right deriv.

    (iii) endpoint values are both zero

  d **(i)** $x = 0$

    (ii) $x = -3$ no left deriv.; $x = -2$ no right deriv.; $x = -1$ no left deriv.; $x = 1$ left deriv. $\neq$ right deriv.; $x = 2\pi$: no right derivative.

    (iii) only 1 endpoint $(-3, 1)$ defined

3 **a** local max at $(2, 13)$; local min at $(4, 9)$

  b local max at $(-1, 16)$; local min at $(3, -16)$

  c local min at $(0, 0)$

  d local max at $(0, 0)$; local mins at $(2, -16)$ and $(-2, -16)$

  e local max at $(0, 0)$; local min at $(1, -3)$

  f no local extrema [horizontal point of inflexion at $(0.5, 0)$]

  g local min at $(2, -166)$ [horizontal point of inflexion at $(-2, 90)$]

  h local max at $(0, 0)$ [point of inflexion at $(3, -27)$]

  i local min at $(0, 0)$ [points of inflexion at $(-1, 1)$ and $(1, 1)$]

Exercise 3 (page 60)

1 **a** global max = 2; global min = -0.25

  b global max = 6; global min = -0.25

  c global max = 27; global min = 0

  d global max = 0.24; global min = -0.25

  e global max = 0.416 29; global min = 0.905 47

  f global max = 1; global min = $1/\sqrt{10}$

2 **a** global max = 2; global min = -1

  b no global max; global min = -1

  c global max = -25; no global min

  d global max = -25; global min = -28

3 **a** global max = 5; global min = 4.0183

  b global max = 1.3863; global min = 0

4 **a** global max = 3.5; no global min

  b no global max; global min = 1.8899

5 **a** global max = 4; global min = -1

  b global max = 1; global min = -10

  c global max = 3; global min = 1

  d global max = 3; global min = -1

  e no global max; global min = -2

  f global max = 1; global min = -1

6 **a** local min at $(1, 0)$; no global max; global min = 0

  b no local max or min; no global max or min

7 **a** endpt max at $(2\pi, 2\pi)$; local max at $(0.90, 0.56)$; local min at $(3.39, -3.29)$

  b global max = 2π; global min = -3.29

8 endpt max at $(2, \ln 16)$; local min at $(0.603, -0.184)$; global max = 2.77

Exercise 4A (page 63)

1 0.96 radians

2 $t = \dfrac{1}{\sqrt{e}}$, $V_{\max} = \dfrac{k}{2e}$

3 $\dfrac{15}{\sqrt{2}}$ feet

4 **b** 400 m

5 drive at the speed limit ($s_{\max}$ = 77 mph)

6 $\tan^{-1} \dfrac{1}{2}$

7 12.5 m

8 **a** $\dfrac{5}{2}$ cm   **b** $\dfrac{5}{2}(1 + \sqrt{3})$

9 radius = 3 cm, height = 3 cm

Exercise 4B (page 64)

1 **a** **(i)** perpendicular   **(ii)** $\dfrac{2}{\sqrt{5}}$

2 35.8 m

3 circle only, radius $\dfrac{F}{2\pi}$

4 $\dfrac{4}{27}\pi h^3 \tan^2 a$

5 half the area of the triangle

Review (page 66)

1 acc = -2 units/sec^2; distance = $8(\ln 2 - 1)$ units

2 global max $x = 18$; global min $x = -0.62$

3 local mins at $(0, 0)$ and $(4, 0)$; local max at $(2, 16)$

4 **a** square has side = $\dfrac{2}{3}L$

  b $1\dfrac{2}{3}L$

CHAPTER 3

Exercise 1A (page 70)

1 $\frac{4}{7}x^7 + C$

2 $x^3 - \frac{7}{2}x^2 + C$

3 $\frac{4}{5}x^5 - 4x^3 + 9x + C$

4 $\frac{4}{3}x^3 - 6x^2 + 9x + C$

5 $\frac{1}{2}x^2 + \frac{7}{x} + C$

6 $\frac{2}{3}x^{\frac{3}{2}} + 14x^{\frac{1}{2}} + C$

7 $\frac{1}{2}\sin 2x + C$

8 $-2\cos 3x + C$

Exercise 1B (page 70)

1 $\frac{1}{32}(8x + 3)^4 + C$

2 $-\frac{1}{24}(5 - 4x)^6 + C$

3 $-\frac{1}{16}(4x - 5)^{-4} + C$

4 $\frac{16}{7}x^7 - 10x^4 + 25x + C$

5 $-\frac{1}{3}\cos(3x - 1) + C$

6 $-\frac{1}{4}\sin(2 - 4x) + C$

7 $-\sin(1 - x) + C$

8 $-\frac{1}{3(3x + 1)} + C$

Exercise 2A (page 72)

1 $\frac{1}{2}e^{2x} + \frac{1}{2}x^2 + C$

2 $\frac{1}{3}\ln|x| + C$

3 $\frac{1}{3}\ln|3x + 2| + C$

4 $\tan x + C$

5 $\frac{5}{2\sqrt{3}} - 1$

6 $e^2(e - 1)$

7 $\frac{1}{2e^2}(e^4 - 4e^3 + 4e^2 + 4e - 1)$

8 $-\frac{3}{4} - \ln 4$

Exercise 2B (page 72)

1 $2 - \frac{\pi}{4}$

2 $\ln\left(\frac{3}{2}\right)$

3 $\frac{\sqrt{3}}{2} - \frac{1}{\sqrt{2}}$

4 $\frac{2}{e} - \frac{1}{2e^2} - \frac{1}{2}$

5 $\frac{1}{6e^6}\left(1 - \frac{1}{e^3}\right)$

6 $\ln 16$

7 $2\pi - 4$

8 $\ln\left(\frac{4}{3}\right)$

Exercise 3 (page 74)

1 $\frac{1}{8}(x^2 + 3)^8 + C$

2 $\frac{1}{3}(x^3 - 4)^5 + C$

3 $\frac{1}{6}(x^2 + 3x)^6 + C$

4 $\frac{1}{8}(4x - x^2)^4 + C$

5 $\frac{1}{6}(3x^2 - 5x)^6 + C$

6 $\frac{3}{5}(x^3 + 2x^2)^5 + C$

7 $-\cos(e^x) + C$

8 $e^{x^2} + C$

9 $\frac{1}{3}e^{x^3+3} + C$

10 $e^{\sin x} + C$

11 $\frac{1}{16}(2x^2 + 1)^4 + C$

12 $-\frac{1}{12}(4x^3 - 2)^{-1} + C$

13 $\frac{3}{14}(x^2 + 2)^7 + C$

14 $\frac{1}{5}\sin^5 x + C$

15 $-\cos^4 x + C$

16 $\frac{1}{5}\tan^5 x + C$

17 $-\frac{4}{5}\cot^5 x + C$

18 $\frac{1}{3}\sec^3 x + C$

19 $-\frac{1}{4}\operatorname{cosec}^4 x + C$

Exercise 4A (page 75)

1 $\frac{2}{3}(x^2 + 4x)^{\frac{3}{2}} + C$

2 $-\frac{1}{3}(x^2 - 1)^{-\frac{3}{2}} + C$

3 $\frac{2}{3}(x^3 - 2x + 1)^{\frac{3}{2}} + C$

4 $\tan^5 x + C$

5 $\frac{1}{4}(\ln|x|)^4 + C$

6 $\frac{1}{6}e^{3x^2+2} + C$

7 $2e^{\sqrt{x}} + C$

8 $-\ln|\cos x| + C$

9 $\frac{1}{3}(a^2 + x^2)^{\frac{3}{2}} + C$

10 $\frac{x}{2}\sqrt{1 - x^2} + \frac{1}{2}\sin^{-1}x + C$

11 $-\frac{1}{16}\cos^4 2x + C$

12 $\frac{2}{3}(x - 1)^{\frac{3}{2}} + 8(x - 1)^{\frac{1}{2}} + C$

Exercise 4B (page 76)

1 $-\cos x + \frac{1}{3}\cos^3 x + C$

2 $\sin x - \frac{1}{3}\sin^3 x + C$

3 $\sin x + \frac{1}{5}\sin^5 x - \frac{2}{3}\sin^3 x + C$

4 $\frac{3}{8}x + \frac{1}{4}\sin 2x + \frac{1}{32}\sin 4x + C$

5 $\frac{3}{8}x - \frac{1}{4}\sin 2x + \frac{1}{32}\sin 4x + C$

6 $\tan x + \frac{1}{3}\tan^3 x + C$

7 $2(1 + \sqrt{x}) - 2\ln|(1 + \sqrt{x})| + C$

8 $\frac{1}{24}(2 + 4x)^{\frac{3}{2}} - \frac{1}{4}(2 + 4x)^{\frac{1}{2}} + C$

9 $\frac{4a + 2bx}{b^2\sqrt{a + bx}} + C$

10 $-\frac{(1 + x^2)^{\frac{3}{2}}}{3x^3} + \frac{\sqrt{1 + x^2}}{x} + C$

11 $\sqrt{1 + x^2} + \frac{1}{\sqrt{1 + x^2}} + C$

12 $\frac{3}{7}(x + 1)^{\frac{7}{3}} - \frac{3}{4}(x + 1)^{\frac{4}{3}} + C$

13 $-\frac{\sqrt{1 + x^2}}{x} + C$

14 $\ln\left|\tan\frac{x}{2}\right| + C$

Exercise 5A (page 77)

1 $\frac{665}{24}$

2 $\frac{1}{2}\ln\left(\frac{92}{17}\right)$

3 $8\ln 2 - 4$

4 $1 - \frac{\sqrt{3}}{2}$

5 $\frac{1}{6}(5\sqrt{5} - 3\sqrt{3})$

6 $\ln\left(\frac{3}{2}\right)$

7 $\ln\left(\frac{\pi}{4}\right)$

8 $\ln\left(\frac{2}{\sqrt{3}}\right)$

9 $\frac{31}{160}$

10 $\sqrt{2} - \frac{2}{\sqrt{3}}$

11 $\ln 2$

12 $\frac{e^2}{2(e^4 - 1)}$

Exercise 5B (page 77)

1 $\frac{8}{15}$

2 $-\frac{7}{60\sqrt{2}} + \frac{2}{15}$

3 a $\frac{8}{u^2 - 4} = \frac{8}{(u+2)(u-2)}$ **b** $2 + 2\ln\left(\frac{5}{3}\right)$

4 a $\frac{1}{u^2 - 1} = \frac{1}{(u+1)(u-1)}$ **b** $\ln\left(\frac{3}{2}\right)$

5 $\frac{\pi}{3} - \frac{\sqrt{3}}{2}$

6 $2\sqrt{3 - \frac{1}{\sqrt{3}}} - 2\sqrt{2}$

Exercise 6A (page 80)

1 a $-\frac{2}{3}(2 - x)^{\frac{3}{2}} + C$ **b** $\frac{1}{5}x^5 + \frac{4}{3}x^3 + 4x + C$

   c $\ln(2 + x^2) + C$ **d** $\frac{1}{2}\ln(1 + e^{2x}) + C$

2 $\frac{19}{3}$ **3** $2\sin\sqrt{x} + C$

4 $-\frac{1}{2(1 + \ln x)^2} + C$ **5** $5\frac{1}{15}$

6 $2\sqrt{x} - 2\ln(1 + \sqrt{x}) + C$ **7** $\frac{31}{15}$

8 $\frac{1}{2}(\ln x)^2 + C$

9 $-\frac{1}{n+1}(1 - \sin x)^{n+1} + C$ **10** $\ln 2$

11 $\frac{2}{3}(2x - 3)^{\frac{3}{2}} + 11(2x - 3)^{\frac{1}{2}} + C$

12 $(\sqrt{x} - 2)^2 + 8(\sqrt{x} - 2) + 8\ln|(\sqrt{x} - 2)| + C$

13 $\frac{47}{480}$

14 $\frac{1}{3}\sec^3 x + C$

15 $-\cos x + \frac{2}{3}\cos^3 x - \frac{1}{5}\cos^5 x + C$

Exercise 6B (page 81)

1 $x + 4\ln(x - 2) + C$

2 $-\frac{1}{3b}\left(\frac{1}{a - b\cos x}\right)^3 + C$

3 $\frac{e^x}{1 + e^x} - \frac{-e^x}{1 - e^x} = \frac{e^x(1 - e^x) - -e^x(1 + e^x)}{(1 + e^x)(1 - e^x)} =$

   $\frac{e^x - e^{2x} + e^x + e^{2x}}{1 - e^{2x}} = \frac{2e^x}{1 - e^{2x}}$

4 $\frac{1}{n+1}(\tan x)^{n+1} + C$

5 $\frac{1}{n+1}$

6 $\frac{1}{2}\ln\left(\frac{1 + \tan x}{1 - \tan x}\right) + C$

7 $\sqrt{2x + 7} + 12 - 12\ln(\sqrt{2x + 7} + 12) + C$

8 $-\frac{1}{4}\ln(3 + 4\cos^2 x) + C$

10 $\frac{5}{72}$ not meaningful

11 $\frac{3}{2} \cdot 7^{\frac{2}{3}} - \frac{3}{2}$

Exercise 7 (page 82)

1 1.45

2 (i) 2.04   **(ii)** 2.185

Exercise 8 (page 84)

See box following exercise

Exercise 9 (page 85)

See box following exercise

Exercise 10A (page 88)

1 (i) $a = 2t$, $x = \frac{1}{3}t^3 + 4t$ **(ii)** $10\frac{2}{3}$ units

2 $v = \frac{1}{2}kt + \frac{1}{4}k\sin 2t + 10$,

   $x = \frac{1}{4}kt^2 + \frac{1}{8}k\cos 2t + 10t$

3 $40\frac{2}{3}$

6 a 14 m/s   **b** $196\frac{2}{3}$ **c** $196\frac{2}{3}$

7 $4\tan 0.5$ units/s, $\ln 4$ units

8 15.708 units

9 $2\sqrt{3} + \frac{\pi}{3}$

10 a $1 - \frac{1}{e^x}$ **b** limit = 1   **c** volume $\to \frac{\pi}{2}$

11 $\frac{4}{3}\pi ab^2$ **12** 2π

13 $6\ln 3 - 4$ **14** 32π

Exercise 10B (page 90)

2 $\left(\frac{1}{\ln 2}, \frac{1}{\ln 16}\right)$ **3** $\frac{1}{5}$

4 $16\frac{2}{3}$ **6** $\frac{1}{3}\pi d^2(3r - d)$

7 a $\frac{\pi}{2}\left(\frac{b - a}{2}\right)^2$

   b semicircle, centre $\left(\frac{a + b}{2}, 0\right)$ radius $\frac{b - a}{2}$

9 b $\frac{\pi}{4\sqrt{2}}$

10 $\frac{9\pi}{8}$

Review (page 93)

1 a $4x^{\frac{1}{2}} - \frac{2}{5}x^{\frac{5}{2}} + C$ **b** $\frac{1}{4}\sin(4x - 1) + C$

   c $\frac{1}{12}(2x + 3)^6 + C$ **d** $\frac{9}{5}x^5 - 2x^3 + x + C$

   e $\frac{1}{3}\tan 3x + C$ **f** $\frac{1}{4}e^{4x-3} + C$

   g $4\ln x + C$ **h** $\frac{1}{2}\tan 2x + C$

2 a $\frac{1}{2}\left(\frac{1}{3}x^3 + 2x\right)^2 + C$ **b** $e^{\cos 2x} + C$

3 $\frac{3}{8}(x^2 + 6x - 1)^{\frac{4}{3}} + C$

4 $\frac{1}{4}\tan^{-1}\left(\frac{x}{4}\right) + C$

7 $4\ln 2$ **9** $\frac{32}{3}$

10 $v_{t=3} = 33$; $x_{t=3} = 36$ **11** $\frac{96\pi}{5}$

CHAPTER 4

Exercise 1 (page 97)

1 a (i) $x \in R$ **(ii)** $f(x) \in [-1, 1]$
   b (i) $x \geq 2$ **(ii)** $f(x) \geq 0$
   c (i) $x \in W$ **(ii)** $f(x) \in W$

   d (i) $x \in R - \left\{\dfrac{(2k - 1)\pi}{2}\right\}$ **(ii)** $f(x) \in R$

   e (i) $x \in R$ **(ii)** $f(x) \geq 0$
   f (i) $x \in R$ **(ii)** $f(x) \in [0, 2]$
   g (i) $x \in R$ **(ii)** $f(x) \geq 0$
   h (i) $x \in R - \{k\pi\}$ **(ii)** $f(x) \geq 1$ or $f(x) \leq -1$
2 a (i) 3   **(ii)** none
   b for each value of y in range $-5 < y < 5$ there are two values of x.
   c $-5 \leq x \leq 5$; $0 \leq y \leq 5$
3 a yes   **b** yes   **c** no
   d no   **e** yes   **f** no

Exercise 2 (page 99)

1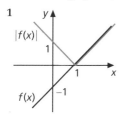

2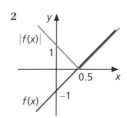

3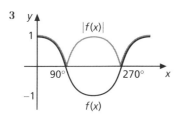

4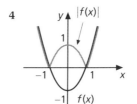

5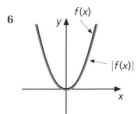

6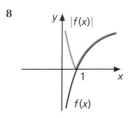

7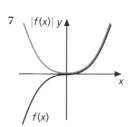

8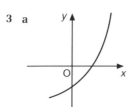

9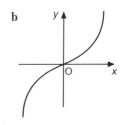

Exercise 3 (page 100)

1 a $\dfrac{x - 4}{3}$ **b** $\dfrac{x + 1}{5}$ **c** $3 - x$

   d $\frac{1}{2}(4 - x)$ **e** $\frac{1}{2}\sqrt[3]{x}$ **f** $\sqrt[5]{1 - x}$

   g $\dfrac{1 - x}{x}$ **h** $\dfrac{x}{x - 1}$ **i** $x^3 + 1$

2 a $\sqrt{x}$; domain $x \geq 0$; range $y \geq 0$

   b $\sqrt{x + 4}$; domain $x \geq 0$; range $y \geq -4$

   c $\sqrt{x} - 1$; domain $x \geq -1$; range $y \geq 0$

   d $\frac{1}{2}(\sqrt{x + 1} + 1)$; domain $x \geq \frac{1}{2}$; range $y \geq -1$

   e $\sqrt{x + 10} - 3$; domain $x \geq -3$; range $y \geq -1$

   f $\sqrt{\dfrac{1 - x}{x}}$; domain $x \geq 0$; range $1 \geq y \geq 0$

   g $\frac{1}{2}(1 + x^2)$; domain $x \geq \frac{1}{2}$; range $y \geq 0$

3 a   **b**

c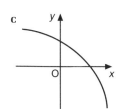

d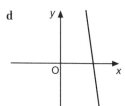

e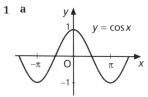

f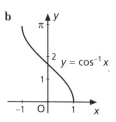

Exercise 4 (page 102)

1 a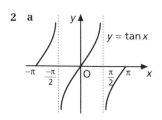

$y = \cos x$

b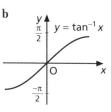

$y = \cos^{-1} x$

2 a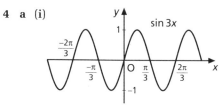

$y = \tan x$

b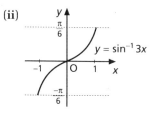

$y = \tan^{-1} x$

3 a domain $0 \le x \le \dfrac{\pi}{2}$; range $y \ge 1$

b domain $x \ge 1$; range $0 \le y \le \dfrac{\pi}{2}$

c domain $0 \le x \le \dfrac{\pi}{2}$; range $y \ge 1$

d domain $x \ge 1$; range $0 \le y \le \dfrac{\pi}{2}$

e domain $0 \le x \le \pi$; range R

f domain R; range $0 \le y \le \pi$

4 a (i)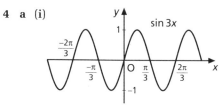

$\sin 3x$

(ii)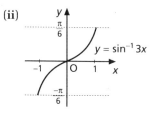

$y = \sin^{-1} 3x$

(iii) $-\dfrac{\pi}{6} \le x \le \dfrac{\pi}{6}$; $-1 \le y \le 1$

b (i)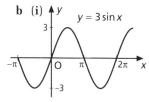

$y = 3\sin x$

(ii)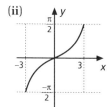

(iii) $-\dfrac{\pi}{2} \le x \le \dfrac{\pi}{2}$; $-3 \le y \le 3$

c (i)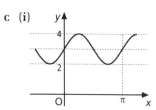

(ii)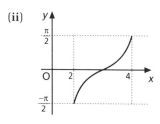

(iii) $-\dfrac{\pi}{2} \le x \le \dfrac{\pi}{2}$; $2 \le y \le 4$

d (i)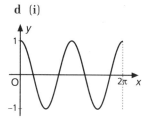

(ii)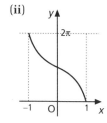

(iii) $0 \le x \le \dfrac{\pi}{2}$; $-1 \le y \le 1$

e (i)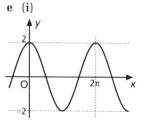

(ii)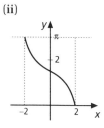

(iii) $0 \le x \le \pi$; $-2 \le y \le 2$

f (i)

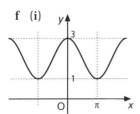

(ii)

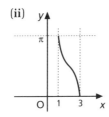

(iii) $0 \le x \le \pi$; $1 \le y \le 3$

g (i)

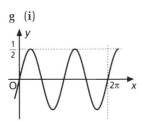

(ii)

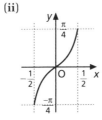

(iii) $-\dfrac{\pi}{4} \le x \le \dfrac{\pi}{4}$; $-\dfrac{1}{2} \le y \le \dfrac{1}{2}$

h (i)

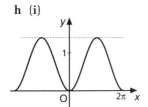

(ii)

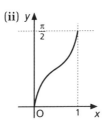

(iii) $0 \le x \le \dfrac{\pi}{2}$; $0 \le y \le 1$

i (i)

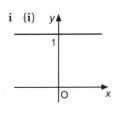

(ii)

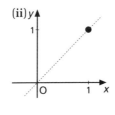

(iii) $x = 1$; $y = 1$

5 a

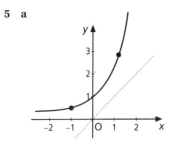

b $x \in R$; $y > 0$

---

c (i)

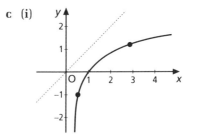

(ii) $\ln x$ (iii) $x > 0$; $y \in R$

6 a (iii) $x \in R$; $y > 0$ **b (iii)** $x \in R$; $y > 0$
c (iii) $x \in R$; $y > -1$ **d (iii)** $x > 0$; $y \in R$
e (iii) $x > 0$; $y \in R$ **f (iii)** $x > 0$; $y \in R$

7 a $x \in R$; $y \ge 2$ **b** $x \in R$; $y \in R$

c $x > 0$; $y < \dfrac{10}{e}$

Exercise 5 (page 103)

1 a min (3, −15)   **b** min (3, −14)
c min (1, −1)
d min (1, −6), max (−2, 21)
e min (6, −108), max (0, 0)
f max (0, 16), min (2, 0), min (−2, 0)

2 a $f(x) = x^4 - 4x^2$

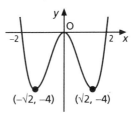

b $f(x) = x^3 - 9x^2$

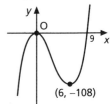
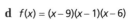

c $f(x) = x^3 - 12x - 16$

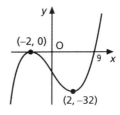

d $f(x) = (x-9)(x-1)(x-6)$

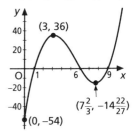

d $f(x) = (x-1)(x-4)^2$

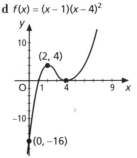

Exercise 6 (page 105)

1 a $(-1, 1)$ end point max;
$(2, 4)$ end point max
b $(-1, 3)$ end point min;
$(1, 3)$ end point min
c $(-3, -30)$ end point min;
$(0, 0)$ end point max
d $(0, 6)$ 0 not in domain;
$(2, -2)$ end point min
e $(-2, 4)$ end point max;
$(3, 6)$ 3 not in domain

2 a $(-1, 0)$ local min; $(3, 16)$ end point max;
$(-2, 1)$ end point max
b $(-1, 0)$ end point min;
$(3, 2)$ end point max;
c $(0, 3)$ local min;
$(-4, 5)$ end point max
d $\left(-\frac{2}{3}, \frac{4}{27}\right)$ local max; $(0, 0)$ local min;
$(2, 12)$ end point max
e $(1, 2)$ local min;
$(3, 3\frac{1}{3})$ end point max

3 a (i)(ii) $(-2, 7)$ endpt max; $(2, -9)$ local min;
$(4, -5)$ endpt max
(iii) $(2, -9)$ global min; $(-2, 7)$ global max,
b (i)(ii) $\left(\frac{\pi}{4}, \frac{1}{\sqrt{2}}\right)$ endpt max; $\left(\frac{\pi}{2}, 0\right)$ local min;
$\left(\frac{2\pi}{3}, \frac{1}{2}\right)$ endpt max
(iii) $\left(\frac{\pi}{2}, 0\right)$ global min; $\left(\frac{\pi}{4}, \frac{1}{\sqrt{2}}\right)$ global max
c (i)(ii) $(1, 0)$ local min; $(e, 1)$ local max
(iii) $(1, 0)$ global min; no global max,
d (i)(ii) $(-1, -11)$ endpt min;
$(2, 16)$ local max; $(3, -27)$ endpt min;
$(0, 0)$ pt of inflexion
(iii) $(3, -27)$ global min;
$(2, 16)$ global max,
e (i)(ii) $\left(-2, -\frac{2}{5}\right)$ endpt max;
$\left(-1, -\frac{1}{2}\right)$ local min
(iii) $\left(-1, -\frac{1}{2}\right)$ global min; no global max

Exercise 7 (page 106)

1 a $f''(x) = 2 > 0$ **b** $f''(x) = -\frac{1}{x^2} < 0$
2 a down   **b** up
c down when $x < 0$; up when $x > 0$;
inflexion at $(0, 0)$
d down when $x < 0$; up when $x > 0$;
inflexion at $(0, 6)$
e (i) up   (ii) down
3 a/b $(0, 0)$ positive, $(\pi, 0)$ negative,
$(2\pi, 0)$ positive

c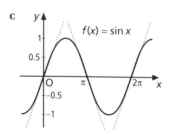

4 a up   **b** up
5 a $(0, 0)$ zero
b $(1, -5)$ negative; $(-1, -5)$ positive
c $(-3, 567)$ negative; $(0, 0)$ zero;
$(3, 567)$ negative
d $\left(\frac{\pi}{2}, 0\right)$ negative
6 $a = \frac{1}{3}(x_1 + x_2 + x_3)$

Exercise 8 (page 108)

1 a $f(x) = 2x^2 + 5$
$f(-x) = 2(-x)^2 + 5$
$f(-x) = 2x^2 + 5$
$\Rightarrow f(x) = f(-x)$
b $f(x) = 3x^5 + 7x^3 - 4x$
$f(-x) = 3(-x)^5 + 7(-x)^3 - 4(-x)$
$f(-x) = -(3x^5 + 7x^3 - 4x)$
$\Rightarrow f(-x) = -f(x)$

2 a even   **b** odd

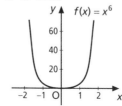

c odd   **d** even

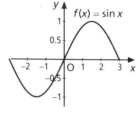
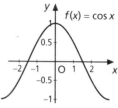

e even   **f** odd

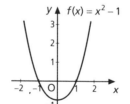
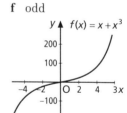

3
a odd	**b** neither	**c** even
d odd	**e** neither	**f** odd
g neither	**h** neither	**i** even
j even	**k** odd	**l** neither

3 $f(x) = \begin{cases} -1 \text{ where } x < 0 \\ \text{undefined at } x = 0 \\ 1 \text{ where } x > 0 \end{cases}$

Exercise 9 (page 109)

1 a (i) $x = -1$ (ii) from left $f(x) \to +\infty$; from right $f(x) \to -\infty$;

b (i) $x = -1$ (ii) from left $f(x) \to -\infty$; from right $f(x) \to +\infty$;

c (i) $x = 1$ (ii) from left $f(x) \to -\infty$; from right $f(x) \to +\infty$;

(i) $x = -2$ (ii) from left $f(x) \to +\infty$; from right $f(x) \to -\infty$;

d (i) $x = 0$ (ii) from left $f(x) \to -\infty$; from right $f(x) \to +\infty$;

(i) $x = -\frac{1}{2}$ (ii) from left $f(x) \to +\infty$; from right $f(x) \to -\infty$;

e (i) $x = 5$ (ii) from left $f(x) \to -\infty$; from right $f(x) \to +\infty$;

(i) $x = -1$ (ii) from left $f(x) \to -\infty$; from right $f(x) \to +\infty$;

f (i) $x = -1$ (ii) from left $f(x) \to -\infty$; from right $f(x) \to +\infty$;

(i) $x = 2$ (ii) from left $f(x) \to -\infty$; from right $f(x) \to +\infty$;

g (i) $x = 1$ (ii) from left $f(x) \to -\infty$; from right $f(x) \to +\infty$;

(i) $x = -1$ (ii) from left $f(x) \to +\infty$; from right $f(x) \to -\infty$;

h (i) $x = 2$ (ii) from left $f(x) \to -\infty$; from right $f(x) \to +\infty$;

(i) $x = 3$ (ii) from left $f(x) \to +\infty$; from right $f(x) \to -\infty$;

i (i) $x = -1$ (ii) from left $f(x) \to -\infty$; from right $f(x) \to +\infty$;

2 a asymptotes $x = (2k - 1)\frac{\pi}{2}$;
from left $f(x) \to +\infty$;
from right $f(x) \to -\infty$ for all k

b asymptotes $x = (2k - 1)\frac{\pi}{2}$;
from left $f(x) \to +\infty$;
from right $f(x) \to -\infty$ for odd k
from left $f(x) \to -\infty$;
from right $f(x) \to +\infty$ for even k

c $x = -1$; from left $f(x) \to -\infty$;
from right $f(x) \to +\infty$

d $x = 0$; from left $f(x) \to +\infty$;
from right $f(x) \to +\infty$

e $x = 1$; from left $f(x) \to -\infty$;
from right $f(x) \to +\infty$

f $x = 0$; from left $f(x) \to -\infty$;
from right $f(x) \to +\infty$.

Exercise 10 (page 110)

1 a (i) $x = 0$
(iii) $y = 0$ (iv) from above as $x \to +\infty$;
from below as $x \to -\infty$.

b (i) $x = 0$ (ii) $f(x) = 1 + \frac{1}{x}$
(iii) $y = 1$ (iv) from above as $x \to +\infty$;
from below as $x \to -\infty$.

c (i) $x = 0$ (ii) $f(x) = x + \frac{2}{x}$
(iii) $y = x$ (iv) from above as $x \to +\infty$;
from below as $x \to -\infty$.

d (i) $x = -1$ (ii) $f(x) = 1 - \frac{2}{x + 1}$
(iii) $y = 1$ (iv) from below as $x \to +\infty$;
from above as $x \to -\infty$.

e (i) $x = -1$ (ii) $f(x) = -1 + \frac{2}{x + 1}$
(iii) $y = -1$ (iv) from above as $x \to +\infty$;
from below as $x \to -\infty$.

f (i) $x = -1$ (ii) $f(x) = x - 1 + \frac{2}{x + 1}$
(iii) $y = x - 1$ (iv) from above as $x \to +\infty$;
from below as $x \to -\infty$.

g (i) $x = 1, x = -1$ (ii) $f(x) = 1 - \frac{2}{x^2 - 1}$
(iii) $y = 1$ (iv) from above as $x \to +\infty$;
from above as $x \to -\infty$.

h (i) none
(iii) $y = 0$ (iv) from above as $x \to +\infty$;
from below as $x \to -\infty$.

i (i) $x = -1, x = 2$
(iii) $y = 0$ (iv) from above as $x \to +\infty$;
from below as $x \to -\infty$.

j (i) $x = 1$ (ii) $f(x) = x + 1 + \frac{4}{x - 1}$
(iii) $y = x + 1$ (iv) from above as $x \to +\infty$;
from below as $x \to -\infty$.

k (i) $x = -2$ (ii) $f(x) = 2x - 2 + \frac{1}{x + 2}$
(iii) $y = 2x - 2$ (iv) from above as $x \to +\infty$;
from below as $x \to -\infty$.

l (i) none (ii) $f(x) = x - \frac{4x}{x^2 + 1}$
(iii) $y = x$ (iv) from below as $x \to +\infty$;
from above as $x \to -\infty$.

2 a (0, 0)   **b** (-1, -1)

3 a $\dfrac{x^3 - x^2 + 1}{x - 1} = x^2 + \dfrac{1}{x - 1} \to x^2$ as $x \to \infty$

b behaves like $y = x^2 + 1$

Exercise 11 (page 112)

1 a $y = \dfrac{1}{x + 3}$

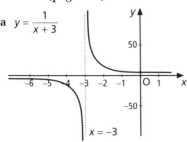

$x = -3$

b $y = \dfrac{3}{2x + 8}$

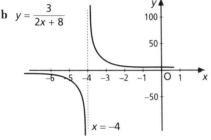

$x = -4$

c $y = \dfrac{x}{x + 2}$

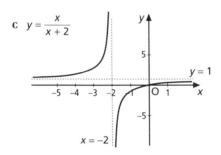

$y = 1$

$x = -2$

d $y = \dfrac{x - 1}{x + 1}$

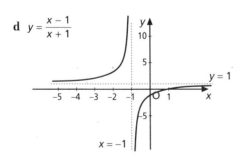

$y = 1$

$x = -1$

e $y = \dfrac{1 - x}{1 + x}$

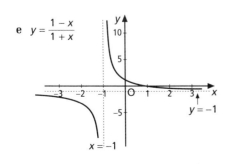

$y = -1$

$x = -1$

f $y = \dfrac{x - 1}{x(x + 1)}$

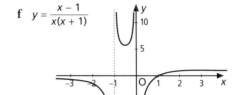

$x = -1$

g $y = \dfrac{x}{(x - 1)(x + 1)}$

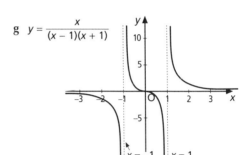

$x = -1$ $x = 1$

h $y = \dfrac{x^2}{x + 1}$

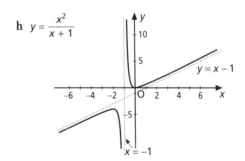

$y = x - 1$

$x = -1$

i $y = x - \dfrac{1}{x}$

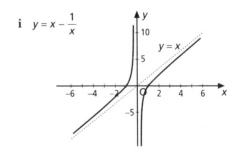

$y = x$

j $y = \dfrac{x^2}{1 - x}$

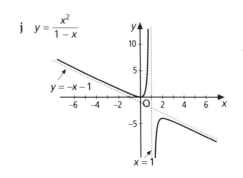

$y = -x - 1$

$x = 1$

155

k $y = \dfrac{(2x + 3)(x - 6)}{(x + 1)(x - 2)}$

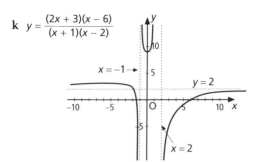

l $y = \dfrac{1}{(x - 2)(x - 4)}$

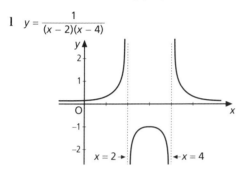

m $y = \dfrac{x^2 - x}{2x + 1}$

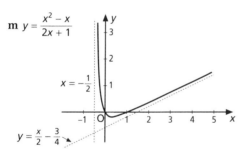

n $y = \dfrac{(x - 1)(x + 2)}{x - 2}$

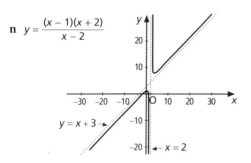

o $y = \dfrac{x^2 - 3x - 10}{x - 2}$

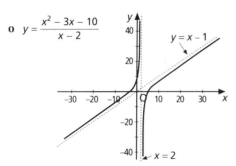

2 a $y = \dfrac{3}{x^2 - 3}$

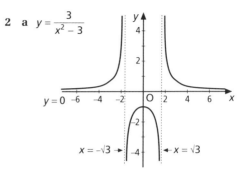

b $y = \dfrac{x^2 - 9}{x}$

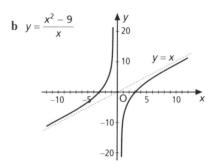

c $y = \dfrac{x^2 + 1}{x^2 - 1}$

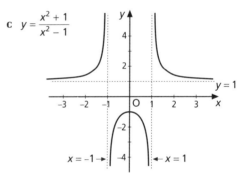

3 b (i) odd

(ii) $f(x) = \dfrac{x}{x^2 + 1}$

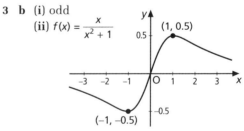

4 a $2x^2 - x + 2 = 2\left[\left(x - \dfrac{1}{4}\right)^2 + \dfrac{15}{16}\right] > 0$

b $a = \dfrac{3}{5},\ b = \dfrac{5}{3}$

c $f(x) = \dfrac{2x^2 + x + 2}{2x^2 - x + 2}$

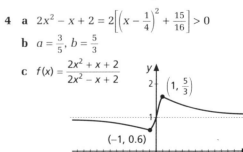

5 **a** vertical asymptote at $x = \frac{5}{3}$ which gives a change of sign.

b $f(x) = \dfrac{3x^2 - 3}{6x - 10}$

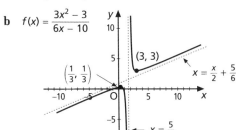

6 **b** $f(x) = \dfrac{2x - 1}{2x^2 - 4x + 1}$

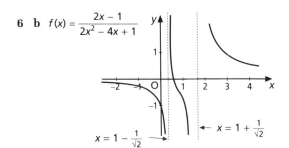

(iv)

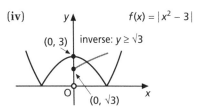

$f(x) = |x^2 - 3|$
inverse: $y \geq \sqrt{3}$

(v)

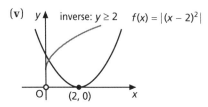

inverse: $y \geq 2$ $f(x) = |(x - 2)^2|$

(vi)

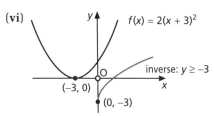

$f(x) = 2(x + 3)^2$
inverse: $y \geq -3$

Exercise 12A (page 114)

1 **a** $y = 2x^2$ **b** $y = -x^2$
 c $y = x^2 + 1$ **d** $y = x^2 - 1$
 e $y = \frac{1}{4}x^2$ **f** $y = (x - 1)^2$
 g $y = (x + 1)^2$ **h** $y = |x^2 - 1|$
 i $y = (x - 2)^2 + 1$ **j** $y = (x + 1)^2 - 1$
 k $y = (x - 1)^2 - 1$ **l** $y = \frac{8}{25}(x - 2)^2 + 1$

2 **a/b** **(i)**

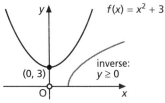

$f(x) = x^2 + 3$
(0, 3)
inverse: $y \geq 0$

(ii)

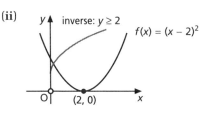

inverse: $y \geq 2$ $f(x) = (x - 2)^2$
(2, 0)

(iii)

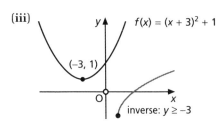

$f(x) = (x + 3)^2 + 1$
(−3, 1)
inverse: $y \geq -3$

3 **a** (0, 0) **b** (0, 2) **c** (0, −2)
 d (−1, 0) **e** (0, 0) **f** (0, 1)
 g none (concavity does not change)

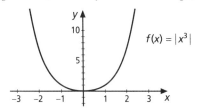

$f(x) = |x^3|$

 h (0, 0) **i** (0, 0) **j** (3, 2)
 k (1, 1) **l** (−1, 0)

4 **a**

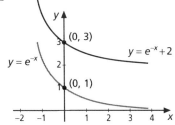

(0, 3)
$y = e^{-x} + 2$
$y = e^{-x}$
(0, 1)

b

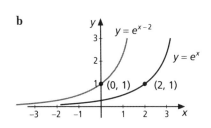

$y = e^{x-2}$
$y = e^x$
(0, 1) (2, 1)

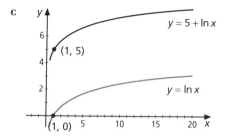

c

$y = 5 + \ln x$

$(1, 5)$

$y = \ln x$

$(1, 0)$

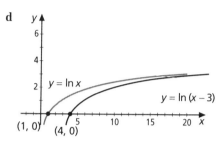

d

$y = \ln x$

$y = \ln (x - 3)$

$(1, 0)$ $(4, 0)$

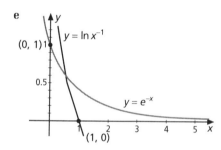

e

$y = \ln x^{-1}$

$(0, 1)$

$y = e^{-x}$

$(1, 0)$

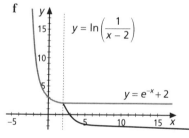

f

$y = \ln\left(\dfrac{1}{x - 2}\right)$

$y = e^{-x} + 2$

5 all sketches are *parabolae* with images of
$(-2, 0)$ and $(0, 1)$ being:

a $(-2, 2), (0, 3)$ **b** $(-2, 0), (0, -1)$
c $(-2, -2), (0, 1)$ **d** $(-4, 0), (-2, 1)$
e $(-2, 0), (0, 2)$ **f** $(-2, 2), (0, 4)$
g $(-4, 0), (0, 1)$ **h** $(-4, 0), (0, 2)$
i $(-4, 2), (0, 4)$

6 transformation and images of $(0, 0)$, $(1, 2)$
and $(2, 0)$ given.
a translation, $(0, 2), (1, 4), (2, 2)$
b translation, $(-2, 0), (-1, 2), (0, 0)$
c reflection (x), $(0, 0), (1, -2), (2, 0)$
d reflection (y), $(0, 0), (-1, 2), (-2, 0)$
e reflect negs, $(0, 0), (1, 2), (2, 0)$
f stretch (y), $(0, 0), (1, 6), (2, 0)$
g squash (x), $(0, 0), (0.5, 2), (1, 0)$

h stretch (y)/squash (x), $(0, 0), (0.5, 6), (1, 0)$
i stretch (y) then translation $(0, 2), (1, 6)$,
$(2, 2)$

7 all sketches are *parabolae* with images of
$(-3, 0)$, $(0, -6)$ and $(3, 0)$ being:
a (i) translation, $(0, 0), (3, -6), (6, 0)$
   (ii) translation, $(-6, 0), (-3, -6), (0, 0)$
   (iii) stretch (y) then translation $(0, 2)$,
    $(1, 6), (2, 2)$
   (iv) reflect (y)/translate, $(-3, 6), (0, 0), (3, 6)$
   (v) stretch (y)/squash (x), $(-1, 0)$,
    $(0, -12), (12, 0)$
   (vi) ...and translate, $\left(-\frac{4}{3}, 0\right), \left(-\frac{1}{3}, -12\right)$,
    $\left(\frac{2}{3}, 0\right)$
   (vii) stretch (y)/reflect (y), $(-2, 0), (1, -12)$,
    $(4, 0)$
   (viii) reflect negs, $(-3, 1), (0, 7), (3, 1)$
b images of $(-3, 2)$, $(0, 0)$ and $(2, -2)$ given.
   (i) $(2, 2), (0, 0), (-3, -2)$
   (ii) $(3, -2), (0, 0), (-2, 2)$
   (iii) $(-3, 6), (-1, 4), (2, 2)$
   (iv) $(2, -3), (0, 0), (-2, 2)$
   (v) $(2, 3), (0, 0), (2, 2)$
   (vi) $(2, 60), (0, 3), (-2, 1)$
c images of $(0, 0)$ $(90, 0)$ $(180, 0)$ $(360, 0)$
given.
   (i) $(0, 0), (-90, 0)$ $(180, 0)$ $(360, 0)$
   (ii) $(0, 1), (-90, 2)$ $(-180, 1)$ $(-360, 0)$
   (iii) $(0, 0), (-90, 2)$ $(-180, 0)$ $(-360, 0)$
   (iv) $(-2, 1), (88, -2)$ $(178, -1)$ $(358, -1)$
   (v) $(0, 0), (90, 0)$ $(180, 0)$ $[(270, 0)]$ $(360, 0)$
   (vi) $(0, 3), (30, 3)$ $(60, 3)$ $(120, 3)$
d images of $(-6, 0)$, $(-4, -5)$ $(0, 0)$ $(4, 5)$
$(6, 0)$ given.
   (i) $(-6, 0), (-4, 5)$ $(0, 0)$ $(4, -5)$ $(6, 0)$
   (ii) $(6, 0), (4, 5)$ $(0, 0)$ $(-4, -5)$ $(-6, 0)$
   (iii) $(-12, 5), (-10, 10)$ $(-6, 5)$ $(-2, 0)$ $(0, 5)$
   (iv) $(0, 0), (2, -5)$ $(6, 0)$ $(10, 5)$ $(12, 0)$
   (v) $(0, 0), (2, 5)$ $(6, 0)$ $(10, 5)$ $(12, 0)$
   (vi) $(-1.5, 3), (-1, 13)$ $(0, 3)$ $(1, -7)$ $(1.5, 3)$
8 a curve symmetrical about y-axis with
points $(-4, 0), (-3, 2), (0, 0), (3, 2)$ $(4, 0)$
b (i) points become $(-4, 0), (-3, -2), (0, 0)$,
$(3, -2)$ $(4, 0)$
   (ii) $(-4, 2), (-3, 4), (0, 2), (3, 4)$ $(4, 2)$
   (iii) $(-8, 2), (-7, 0), (-4, 2), (-1, 0)$ $(0, 2)$
   (iv) $(-4, 0), (-3, -2), (0, 0), (3, -2)$ $(4, 0)$

Exercise 12B (page 116)

1 asymptote $y = 0$ and points $(-1, -1)$ $(1, 1)$
become
a $y = 0$, $(-1, 1)$ $(1, -1)$
b $y = 1$, $(-1, -1)$ $(1, 3)$
c $y = 2$, $(-1, 3)$ $(1, 1)$

d $y = 0$, $(-1, 1)$ $(1, 1)$ see sketch

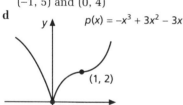

$$y = \frac{1}{|x|}$$

2 a parabola concave up passing through $(-1, 2)$ and $(0, 3)$

b parabola concave up passing through $(-3, -10)$ and $(0, -1)$

c parabola concave down passing through $(-1, 0)$, $(0, 3)$, $(1, 4)$ $(3, 0)$

d as graph **b** but *tails* reflected in y-axis

e as graph **c** but *tails* reflected in y-axis

f parabola concave down passing through $(0, 0)$, $(4, -16)$, $(8, 0)$

3 a similar profile passing through $(0, 1)$ and $(1, 2)$

b profile reflected in y-axis passing through $(0, 0)$ and $(-1, 1)$

c profile reflected in y-axis passing through $(-1, 5)$ and $(0, 4)$

d

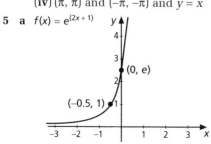

$$p(x) = -x^3 + 3x^2 - 3x$$

e profile reflected in x-axis passing through $(0, 0)$ and $(1, -1)$

f profile reflected in $x = y$ passing through $(0, 0)$ and $(1, 1)$

4 a $f'(x) \geq 0$ $\forall x$

b end points (π, π) and $(-\pi, -\pi)$ and $y = x$ become

  (i) $\left(\frac{\pi}{2}, \pi\right)$ and $\left(-\frac{\pi}{2}, -\pi\right)$ and $y = 2x$

  (ii) $(-\pi, \pi)$ and $(\pi, -\pi)$ and $y = -x$

  (iii) $(-2\pi, -\pi)$ and (0π) and $y = x + \pi$

  (iv) (π, π) and $(-\pi, -\pi)$ and $y = x$

5 a $f(x) = e^{(2x+1)}$

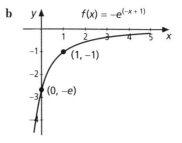

b

$$f(x) = -e^{(-x+1)}$$

c

$$f(x) = |5 - e^{-x}|$$

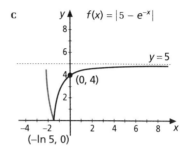

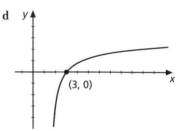

d

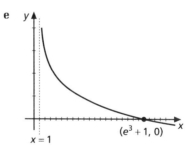

e

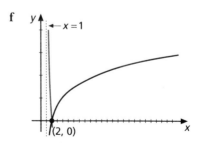

f

6 • zeroes become vertical asymptotes and vice versa.

 • horizontal asymptotes are shared.

 • maxima become minima and vice versa.

Review (page 117)

1 domain: $x \in W$ range: subset of the natural numbers defined by the recurrence relation $\{u_n : u_n = nu_{n-1}; u_0 = 1, n \in W\}$

2 **a** $(1.5, -1.5), (2, -1)$
   b/c $(1.5, -1.5)$ global min, $(2, -1)$ endpt max, no local or global max

3 $x \geq -3$ concave up; $x \leq -3$ concave down; $(-3, 1)$ inflexion.

4 **a**

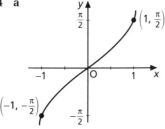

   b Domain: $-1 \leq x \leq 1$; range $-\dfrac{\pi}{2} \leq y \leq \dfrac{\pi}{2}$

5 **a** $f(x) = \dfrac{x-2}{x+1}$

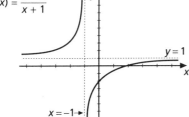

   b $f(x) = \dfrac{x+3}{x^2 + x - 2}$

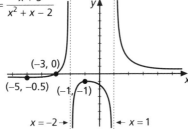

   c $f(x) = \dfrac{x^2}{x-3}$

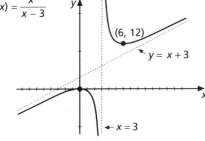

d $f(x) = \dfrac{(x-3)(x+6)}{(x-1)(x+2)}$

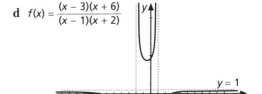

e $f(x) = \dfrac{x^2 + 2x - 3}{x+1}$

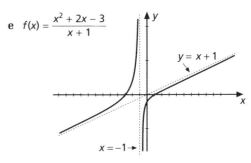

6 cubic sketches where the points $(0, 4)$, $(3, 0)$, $(6, 0)$, $(9, 0)$ become:
   a $(0, 8), (3, 0), (6, 0), (9, 0)$
   b $(0, 4), (-3, 0), (-6, 0), (-9, 0)$
   c $(0, -4), (3, 0), (6, 0), (9, 0)$
   d $(0, 2), (3, -2), (6, -2), (9, -2)$
   e $(-4, 4), (-1, 0), (2, 0), (5, 0)$
   f $(0, -4), (1, 0), (2, 0), (3, 0)$
   g $(-2, 7), (-1, -1), (0, -1), (1, -1)$
   h

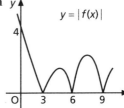

7 **a**

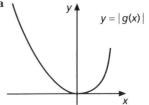

   b $y = h(x)$
   (where $h^{-1}(x) = g(x)$)

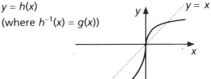

160

8 **a** $a > 0$

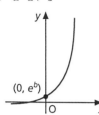

(0, e^b)

b $a > 0, b < 0$

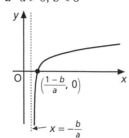

$\left(\dfrac{1-b}{a}, 0\right)$

$x = -\dfrac{b}{a}$

c $x + y = 7$
$x = 4$

d $x = 5$
$y = 2$

CHAPTER 5

Exercise 1 (page 121)

1 **a** $x = 1, y = 4$ **b** $x = -\frac{1}{3}, y = 3\frac{2}{9}$
 c redundant $(x, 3 - 4x)$ **d** inconsistent
 e $x = 2, y = -1$ **f** $x = 0, y = 2$
2 **a** redundant $(3, 1)$ **b** inconsistent
 c inconsistent
 d redundant $(-1, -1)$
3 Two of the equations in the system are
 redundant. Solution is $\left(x, \dfrac{5 - 2x}{3}\right)$
4 **a** $c + 3t = 3.6$; $2c + 2t = 4$; $3c + 2t = 5$
 b $c = £1.20$; $t = £0.80$
 c Inconsistent. Using solution, price
 should be £5.20 not £5.

Exercise 2 (page 123)

1 **a** $\begin{pmatrix} 1 & 2 \\ 3 & -1 \end{pmatrix}\begin{pmatrix} x \\ y \end{pmatrix} = \begin{pmatrix} 4 \\ 5 \end{pmatrix}$ **b** $\begin{pmatrix} 2 & 4 \\ 1 & -2 \end{pmatrix}\begin{pmatrix} x \\ y \end{pmatrix} = \begin{pmatrix} 8 \\ -4 \end{pmatrix}$

$\left(\begin{array}{cc|c} 1 & 2 & 4 \\ 3 & -1 & 5 \end{array}\right)$ $\left(\begin{array}{cc|c} 2 & 4 & 8 \\ 1 & -2 & -4 \end{array}\right)$

c $\begin{pmatrix} 3 & -2 \\ 1 & -1 \end{pmatrix}\begin{pmatrix} x \\ y \end{pmatrix} = \begin{pmatrix} 7 \\ 2 \end{pmatrix}$ **d** $\begin{pmatrix} 3 & 0 \\ 4 & 1 \end{pmatrix}\begin{pmatrix} x \\ y \end{pmatrix} = \begin{pmatrix} -6 \\ -7 \end{pmatrix}$

$\left(\begin{array}{cc|c} 3 & -2 & 7 \\ 1 & -1 & 2 \end{array}\right)$ $\left(\begin{array}{cc|c} 3 & 0 & -6 \\ 4 & 1 & -7 \end{array}\right)$

e $\begin{pmatrix} 0.5 & 2 \\ 1 & -1.5 \end{pmatrix}\begin{pmatrix} x \\ y \end{pmatrix} = \begin{pmatrix} 3 \\ -5 \end{pmatrix}$

$\left(\begin{array}{cc|c} 0.5 & 2 & 3 \\ 1 & -1.5 & -5 \end{array}\right)$

f $\begin{pmatrix} 0.4 & 0.6 \\ 0.1 & -0.2 \end{pmatrix}\begin{pmatrix} x \\ y \end{pmatrix} = \begin{pmatrix} 0.2 \\ 0.4 \end{pmatrix}$

$\left(\begin{array}{cc|c} 0.4 & 0.6 & 0.2 \\ 0.1 & -0.2 & 0.4 \end{array}\right)$

2 **a** $2x + y = 3$ **b** $x + 2y = 0$
 $3x + 4y = 7$ $3x + y = 5$

Exercise 3 (page 124)

1 **a** (i) $\left(\begin{array}{cc|c} 2 & 1 & 6 \\ 1 & -1 & 1 \end{array}\right)$ (ii) $\left(\begin{array}{cc|c} 1 & -1 & 1 \\ 0 & 3 & 4 \end{array}\right)$

 (iii) $x - y = 1$ $x = 2\frac{1}{3}$
  $3y = 4$ $y = 1\frac{1}{3}$

 b (i) $\left(\begin{array}{cc|c} 2 & 3 & -1 \\ 1 & -2 & -4 \end{array}\right)$ (ii) $\left(\begin{array}{cc|c} 1 & -2 & -4 \\ 0 & 7 & 7 \end{array}\right)$

 (iii) $x - 2y = -4$ $x = -2$
  $7y = 7$ $y = 1$

 c (i) $\left(\begin{array}{cc|c} 4 & -3 & 22 \\ 2 & 5 & -2 \end{array}\right)$ (ii) $\left(\begin{array}{cc|c} 2 & 5 & -2 \\ 0 & -13 & 26 \end{array}\right)$

 (iii) $2x + 5y = -2$ $x = 4$
  $-13y = 26$ $y = -2$

 d (i) $\left(\begin{array}{cc|c} 2 & -1 & -1 \\ 3 & -5 & -4 \end{array}\right)$ (ii) $\left(\begin{array}{cc|c} 2 & -1 & -1 \\ 0 & -7 & -5 \end{array}\right)$

 (iii) $2x - y = -1$ $x = -\frac{1}{7}$
  $-7y = -5$ $y = \frac{5}{7}$

2 **a** (i) $\begin{pmatrix} 3 & -5 & -8 \\ 2 & -1 & -3 \end{pmatrix}$ (ii) $\begin{pmatrix} 1 & 0 & -1 \\ 0 & 1 & 1 \end{pmatrix}$

 (iii) $x = -1$
  $y = 1$

 b (i) $\begin{pmatrix} 4 & 7 & 5 \\ 3 & -4 & 13 \end{pmatrix}$ (ii) $\begin{pmatrix} 1 & 0 & 3 \\ 0 & 1 & -1 \end{pmatrix}$

 (iii) $x = 3$
  $y = -1$

 c (i) $\begin{pmatrix} 2 & -5 & -4 \\ 5 & -2 & -10 \end{pmatrix}$ (ii) $\begin{pmatrix} 1 & 0 & -2 \\ 0 & 1 & 0 \end{pmatrix}$

 (iii) $x = -2$
  $y = 0$

 d (i) $\begin{pmatrix} 6 & -4 & -36 \\ 9 & 2 & -6 \end{pmatrix}$ (ii) $\begin{pmatrix} 1 & 0 & -2 \\ 0 & 1 & 6 \end{pmatrix}$

 (iii) $x = -2$
  $y = 6$

3 **a** (i) $\left(\begin{array}{cc|c} 2 & -1 & 1 \\ 6 & -3 & 3 \end{array}\right)$ (ii) $\left(\begin{array}{cc|c} 2 & -1 & 1 \\ 0 & 0 & 0 \end{array}\right)$

 (iii) zero row $\Rightarrow$ redundancy

 b (i) $\left(\begin{array}{cc|c} 1 & 3 & 7 \\ 4 & 12 & 28 \end{array}\right)$ (ii) $\left(\begin{array}{cc|c} 1 & 3 & 7 \\ 0 & 0 & 0 \end{array}\right)$

 (iii) as **a**

4 a (i) $\begin{pmatrix} 3 & -2 & | & 1 \\ 6 & -4 & | & 3 \end{pmatrix}$ **(ii)** $\begin{pmatrix} 3 & -2 & 1 \\ 0 & 0 & 1 \end{pmatrix}$

(iii) $0 = a \ (a \neq 0) \Rightarrow$ inconsistency

b (i) $\begin{pmatrix} 4 & 2 & | & 5 \\ 2 & 1 & | & 3 \end{pmatrix}$ **(ii)** $\begin{pmatrix} 4 & 2 & | & 5 \\ 0 & 0 & | & 1 \end{pmatrix}$

(iii) as **a**

5 a $9 = a - b$; $12 = 4a - 16b$

b $\begin{pmatrix} 1 & -1 & | & 9 \\ 4 & -16 & | & 12 \end{pmatrix}$ $a = 11$; $b = 2$

c $\frac{11}{2}$ units

6 a $\begin{pmatrix} 3 & 1 & | & 10 \\ 5 & -2 & | & -9 \end{pmatrix}$; $(1, 7)$

b (i) redundant **(ii)** inconsistent
c (i) redundancy $\Rightarrow$ same line
(ii) inconsistency $\Rightarrow$ parallel lines.

Exercise 4A (page 127)

1 a (i) $\begin{pmatrix} 1 & 2 & 1 & | & 8 \\ 3 & 1 & -2 & | & -1 \\ 1 & 5 & -1 & | & 8 \end{pmatrix}$ **(ii)** $\begin{pmatrix} 1 & 2 & 1 & | & 8 \\ 0 & -5 & -5 & | & -25 \\ 0 & 0 & -5 & | & -15 \end{pmatrix}$

(iii) $x = 1$
$y = 2$
$z = 3$

b (i) $\begin{pmatrix} 2 & 3 & -1 & | & -1 \\ 1 & -3 & -2 & | & 4 \\ 5 & 1 & 3 & | & 4 \end{pmatrix}$ **(ii)** $\begin{pmatrix} 1 & -3 & -2 & | & 4 \\ 0 & 9 & 3 & | & -9 \\ 0 & 0 & -\frac{7}{3} & | & 0 \end{pmatrix}$

(iii) $x = 1$
$y = -1$
$z = 0$

c (i) $\begin{pmatrix} 3 & 1 & 0 & | & 5 \\ 1 & 2 & -3 & | & -12 \\ 1 & 0 & 2 & | & 10 \end{pmatrix}$ **(ii)** $\begin{pmatrix} 1 & 0 & 2 & | & 10 \\ 0 & 1 & -6 & | & -25 \\ 0 & 0 & 7 & | & 28 \end{pmatrix}$

(iii) $x = 2$
$y = -1$
$z = 4$

d (i) $\begin{pmatrix} 3 & -4 & 1 & | & 24 \\ 1 & -2 & -2 & | & 7 \\ 1 & 1 & 1 & | & 4 \end{pmatrix}$ **(ii)** $\begin{pmatrix} 1 & 1 & 1 & | & 4 \\ 0 & -3 & -3 & | & 3 \\ 0 & 0 & -5 & | & 5 \end{pmatrix}$

(iii) $x = 5$
$y = -2$
$z = 1$

e (i) $\begin{pmatrix} 4 & 2 & 1 & | & 3 \\ 1 & 3 & 5 & | & 3 \\ 2 & 0 & 3 & | & 5 \end{pmatrix}$ **(ii)** $\begin{pmatrix} 1 & 3 & 5 & | & 3 \\ 0 & -10 & -19 & | & -9 \\ 0 & 0 & 1 & | & 1 \end{pmatrix}$

(iii) $x = 1$
$y = -1$
$z = 1$

f (i) $\begin{pmatrix} 1 & 1 & 5 & | & 0 \\ 4 & 1 & -6 & | & -17 \\ 1 & -1 & -1 & | & 0 \end{pmatrix}$ **(ii)** $\begin{pmatrix} 1 & 1 & 5 & | & 0 \\ 0 & 1 & 3 & | & 0 \\ 0 & 0 & 17 & | & 17 \end{pmatrix}$

(iii) $x = -2$
$y = -3$
$z = 1$

2 a $a + b + c = 2$, $4a + 2b + c = 7$,
$9a + 3b + c = 14$
b $a = 1$, $b = 2$, $c = -1$
c $y = x^2 + 2x - 1$

3 a $-4g - 2f + c = -5$, $-2g + 4f + c = -5$,
$12g + 6f + c = -45$
b $g = -3$; $f = 1$, $c = -15$;
$x^2 + y^2 - 6x + 2y - 15 = 0$
c $r = 5$

4 a $s + c + g = 185$, $3s + 4c + 2g = 460$,
$2s + 3c + 2g = 375$
$s = 80$, $c = 5$, $g = 100$
b After 1 hr there is 5 seconds of GO phase
left.

Exercise 4B (page 128)

1 a $x = 1$, $y = 1$, $z = 1$ **b** $x = 1$, $y = -2$, $z = 1$
c $x = -1$, $y = 3$, $z = 0$ **d** $x = 4$, $y = 2$, $z = 3$
e $x = 5$, $y = 1$, $z = 1$ **f** $x = 2$, $y = -3$, $z = 2$
2 a $a + b - c = -3$, $a + 2b - 4c = -28$,
$a - b - c = -13$
b $a = 2$, $b = 5$, $c = 10$
c $y = 2 + 5x - 10x^2$

Exercise 5 (page 130)

1 a redundant; $x = 9z + 10$, $y = -16z - 15$, $z = z$
b $x = 1$, $y = 2$, $z = -1$
c inconsistent – no solutions
d $x = 3$, $y = 1$, $z = -1$
e inconsistent – no solutions
f x, $\dfrac{17x + 33}{8}$, $\dfrac{-11x - 3}{8}$

2 $k = 0$
3 $k = 9$
4 a $d = -9$ and $e \neq 1$
b $d = -9$ and $e = 1$ **c** $d \neq -9$

Exercise 6 (page 132)

1 a 1, 1, 2, 1 **b** 1, 2, 0, -1
c 1, 5, 6, 2 **d** 1, 3, 5, 1
2 a $a = \frac{1}{2}$, $b = \frac{1}{2}$, $c = 0$; $S_n = \frac{1}{2}n^2 + \frac{1}{2}n$

b $a = \frac{1}{3}$, $b = \frac{1}{2}$, $c = \frac{1}{6}$, $d = 0$;
$S_n = \frac{1}{3}n^3 + \frac{1}{2}n^2 + \frac{1}{6}n$

c $a = \frac{1}{4}$, $b = \frac{1}{2}$, $c = \frac{1}{4}$, $d = 0$;
$S_n = \frac{1}{4}n^4 + \frac{1}{2}n^3 + \frac{1}{4}n^2$

3 a $\begin{pmatrix} 0.500 & 0.866 & 1 & | & 11.0 \\ 0.985 & -0.174 & 1 & | & 8.26 \\ 0.866 & -0.500 & 1 & | & 6.60 \end{pmatrix}$

b $a = 2.96, b = 4.01, c = 6.05$;
$y = 3 \sin x + 4 \cos x + 6$

c $1 < y < 11$

Exercise 7 (page 133)

1 a (i) $[11.5, 12.5]$ **(ii)** 0.5 **(iii)** 4%
  b (i) $[24.5, 25.5]$ **(ii)** 0.5 **(iii)** 2%
  c (i) $[45.5, 46.5]$ **(ii)** 0.5 **(iii)** 1%
  d (i) $[2.35, 2.45]$ **(ii)** 0.05 **(iii)** 2%
  e (i) $[7.65, 7.75]$ **(ii)** 0.05 **(iii)** 0.6%
  f (i) $[9.05, 9.15]$ **(ii)** 0.05 **(iii)** 0.5%
  g (i) $[0.445, 0.455]$ **(ii)** 0.005 **(iii)** 1%
  h (i) $[0.355, 0.365]$ **(ii)** 0.005 **(iii)** 1%
  i (i) $[0.025, 0.035]$ **(ii)** 0.005 **(iii)** 17%

2 a (i) sum 35 ± 1; diff 21 ± 1
    (ii) 1 **(iii)** $3\%, 5\%$
  b (i) sum 155 ± 1; diff 1 ± 1
    (ii) 1 **(iii)** $0.6\%, 100\%$
  c (i) sum 12.7 ± 0.1; diff 0.1 ± 0.1
    (ii) 0.1 **(iii)** $0.8\%, 100\%$
  d (i) sum 0.84 ± 0.01; diff 0.78 ± 0.01
    (ii) 0.01 **(iii)** $1\%, 1\%$
  e (i) sum 2.25 ± 0.01; diff 0.25 ± 0.01
    (ii) 0.01 **(iii)** $0.4\%, 4\%$
  f (i) sum 0.021 ± 0.001; diff 0.001 ± 0.001
    (ii) 0.001 **(iii)** $5\%, 100\%$

3 a 46 ± 1 **b** 2.3 ± 0.1 **c** 1 ± 1
  d 85 ± 1 **e** 2 ± 1 **f** 0.1 ± 0.1

Exercise 8 (page 136)

1 a (i)

	34.5	35	35.5
16.5	(3, 1.5)	(2.17, 2)	(1.33, 2.5)
17	(4.83, 0.5)	(4, 1)	(3.17, 1.5)
17.5	(6.67, −0.5)	(5.83, 0)	(5, 0.5)

(ii)-(iii)
    $1.33 \le x \le 6.67$ $x = 4 \pm 2.67$ (67%)
    $-0.5 \le y \le 2.5$ $y = 1 \pm 1.5$ (150%)
(iv) ill-conditioned.

b (i)

	3.5	4	4.5
12.5	(0.77, 1.95)	(1.05, 1.91)	(1.32, 1.86)
13	(0.73, 2.05)	(1, 2)	(1.27, 1.95)
13.5	(0.68, 2.14)	(0.95, 2.09)	(1.23, 2.05)

(ii)-(iii)
    $0.68 \le x \le 1.32$ $x = 1 \pm 0.32$ (32%)
    $1.86 \le y \le 2.14$ $y = 2 \pm 0.14$ (7%)
(iv) Poor, but not ill-conditioned.

c (i)

	18.5	19	19.5
25.5	(2.5, 2)	(0.5, 5.5)	(−1.5, 9)
26	(4, −0.5)	(2, 3)	(0, 6.5)
26.5	(5.5, −3)	(3.5, 0.5)	(1.5, 4)

(ii)-(iii)
    $-1.5 \le x \le 5.5$ $x = 2 \pm 3.5$ (175%)
    $-3 \le y \le 9$ $y = 3 \pm 6$ (200%)
(iv) ill-conditioned

2 a no **b** yes **c** yes **d** yes

3 a $10x + 10y = 1700, 10x + 11y = 1790$
  b $x = 80, y = 90$
  c using 1700.5 and 1789.5 (81.05, 89);
    using 1699.5 and 1790.5 (78.95, 91)
    $x = 80 \pm 1.05, y = 90 \pm 1$ (1% error)

4 ill-conditioned.
  At least $x = 8 \pm 11, y = 4 \pm 1.5, z = 7 \pm 9$

5 Solution 0, 0, 1, −1.
  Using 2, 1, 1, 1 instead gives 1, −1, −1, 1

Review (page 138)

1 $x = 1, y = -1$

2 a (i) $\begin{pmatrix} 2 & 1 & 1 \\ 3 & 2 & -1 \\ 1 & -1 & 0 \end{pmatrix} \begin{pmatrix} x \\ y \\ z \end{pmatrix} = \begin{pmatrix} 2 \\ 6 \\ 0 \end{pmatrix}$

  (ii) $\begin{pmatrix} 2 & 1 & 1 & | & 2 \\ 3 & 2 & -1 & | & 6 \\ 1 & -1 & 0 & | & 0 \end{pmatrix}$

  b $\begin{pmatrix} 1 & -1 & 0 & | & 0 \\ 0 & 3 & 1 & | & 2 \\ 0 & 5 & -1 & | & 6 \end{pmatrix} \rightarrow \begin{pmatrix} 1 & -1 & 0 & | & 0 \\ 0 & 3 & 1 & | & 2 \\ 0 & 0 & -1 & | & 1 \end{pmatrix}$

  c $x = 1, y = 1, z = -1$

3 $x = 5, y = -1, z = -3$

4 a $9a + b + 3c = 7, a + b - c = 7,$
    $a + 4b + c = 15$
  b $a = 1, b = 4, c = -2$
  c $x^2 + 4y^2 - 2x - 8y + 1 = 0$

5 a $\begin{pmatrix} 1 & 0 & 0 & | & 4 \\ 0 & 1 & 0 & | & 1 \\ 0 & 0 & 1 & | & -3 \end{pmatrix}$ **b** $x = 4, y = 1, z = -3$

6 a $p = 1$ and $q \ne \frac{4}{3}$
  b $p = 1$ and $q = \frac{4}{3}$

7 (i) is ill-conditioned.

Index